AF415882

Structural Analysis II
Lecture Notes

Peter I. Kattan

Petra Books
www.PetraBooks.com

Peter I. Kattan, PhD

Correspondence about this book may be sent to the author at one of the following two email addresses:

pkattan@petrabooks.com

info@petrabooks.com

Structural Analysis II Lecture Notes
written by Peter I. Kattan.
ISBN: 979-8-8690-3319-2
All rights reserved. No part of this book may be copied or reproduced without written permission of the author or publisher.
© 2023 Peter I. Kattan

In Loving Memory of My Parents

Structural Analysis II Lecture Notes

Structural Analysis II Lecture Notes

Preface

These are the handwritten notes for the course "Structural Analysis II" that was taught at Applied Science University in the period 1996-1998. These notes are based on the book "Structural Analysis" by Alexander Chajes, Second Edition. This book is out of print at the present time. Sutdents find these notes useful and helpful in their studies and the author is glad to make them available to students worldwide. It is hoped that these notes will also revive the Chajes book.

The Table of Contents lists details of all the chapters but providing no page numbers. The specific needed chapter can be found easily. A chapter on matrix methods (The Flexibility Matrix Method) is included to introduce the student to matrix methods but it is not the aim of these notes to teach matrix structural analysis. Therefore, The Stiffness Matrix Method is not included.

December 2023   Peter I. Kattan

Contents

History of Structural Analysis :

① Tremendous progress in structural analysis occurred during the latter part of the 19th century and the early part of the twentieth century.

    1. Mohr / Germany
    2. Muller-Breslau / Germany
    3. Maxwell / England
    4. Castigliano / Italy
    5. Greene / America (U.S.A.)
    7. Hardy Cross

} Routine techniques for the analysis of **both** determinate and indeterminate structures.

② **1856** , Henry Bessemer / England developed the first successful process for producing large quantities of _steel_ economically.

Earlier contributors to Structural Analysis :

1. Galileo (1564 – 1642) laid the ground work for the science of structural analysis.

2. Robert Hooke (1635 – 1703) formulated the law of linear behavior of materials -- Hooke's Law.

3. Sir Issac Newton (1642 – 1727) formulated the laws of motion.
    1687: Principia (Mathematical Principles of Natural Philosophy)

4. Leonhard Euler (1707 – 1783) : the first to analyze correctly the buckling of columns.

As a result of two factors :
    (1) availability of steel.
    (2) indeterminate structural analysis

we have new types of structures :

    * Long-span Suspension Bridges.
    * High-rise Buildings (Skyscrapers).

Cable Structures / Suspension Bridges:

. Brooklyn Bridge of New York City   (1883):
    the first suspension bridge to use steel-wire cable.
2. Golden Gate Bridge of San Francisco (1937).

History of Indeterminate Structural Analysis:

① 1717,   John Bernoulli ⟹   Method of Virtual Work.

② 1864, James Clerk Maxwell
        ⟹   Maxwell's Law of Reciprocal Deflections

③ 1872, E. Betti ⟹   Betti's Law (for deflections).

④ 1879, Carlo Alberto Castigliano ⟹ Castigliano's Theorems.

① 1864, James Clerk Maxwell ⟹ Method of Consistent Deformations
  1874, Otto Mohr   (Maxwell-Mohr Method)
  1882 (Mohr's Circle)   (General Method)

② 1858, Ménabréa ⟹ Theorem (Method) of Least Work
   1875, Castigliano (at Turin)
       (published in 1879).

③ 1880 Heinrich Manderla (for trusses)(limited)
   1892 Otto Mohr (for trusses)(limited). Slope-Deflection Method
   1914 Alex Bendixen (general/frames)
   1915 George Maney (University of Minnesota)   ⟸ (independently)

④ Professor Hardy Cross, 1924 (to students
        at the University of Illinois)   Moment Distribution Method
   1930
   1932) publication of the method
   1932 C.T. Morris : Sidesway Correction for Moment Distribution
⑤ 1932, C.T. Morris : Sidesway Correction for Moment Distribution
⑥ Late 1950s, 1960s (matrix/computer methods), 1970s, ⟹ 1980s P.C.'s.

1886, Professor Heinrich Müller-Breslau
        ⟹ Müller-Breslau Principle for Influence Lines.

⑦ 1873, Charles E. Greene (University of Michigan) ⟹ Moment Area Theorems.
⑧ 1860 Otto Mohr ⟶ Elastic Weight Method ⟹ Conjugate-Beams Method.

<u>Methods of Indeterminate Structural Analysis:</u>

<u>Type 1</u> : <u>Force Methods</u> :

If the forces (reactions) are selected as the unknowns, then force methods are used.

In this case, the equations of equilibrium are not sufficient, and additional equations must be formulated based on compatibility (consistency) of displacements.

Force methods are also called <u>compatibility methods</u>.

<u>Examples</u> :
1. Method of Consistent Deformations.
2. Method of Least Work.
3. Three-Moment Equation Method.
4. Column Analogy Method.
5. Flexibility Matrix Method.
6. Elastic Center Method.

<u>Type 2</u> : <u>Displacement Methods:</u>

If the displacements (deflections) are selected as the unknowns, then displacement methods are used.

Displacement methods are also called <u>equilibrium methods</u>.

<u>Examples</u> :
1. Slope-Deflection Method.
2. Moment Distribution Method.
3. Stiffness Matrix Method.

Methods of indeterminate structural analysis can also be classified as.

(1) <u>Classical Methods</u> : These are old methods performed manually.

(Method of Consistent Deformations, Method of Least Work, Slope-Deflection Method, Moment Distribution Method).

(2) <u>Modern Methods:</u> (Matrix Methods):

these are methods that are programmed easily using the computer.

(Flexibility Matrix Method, Stiffness Matrix Method).

Maxwell's Law of Reciprocal Deflections:

> the deflection at A due to a load applied at B is equal to the deflection at B if the same load is applied at A.

$$\delta_{AB} = \delta_{BA}$$

where

δ_{AB} : deflection at A due to a load at B.

δ_{BA} : deflection at B due to a load at A.

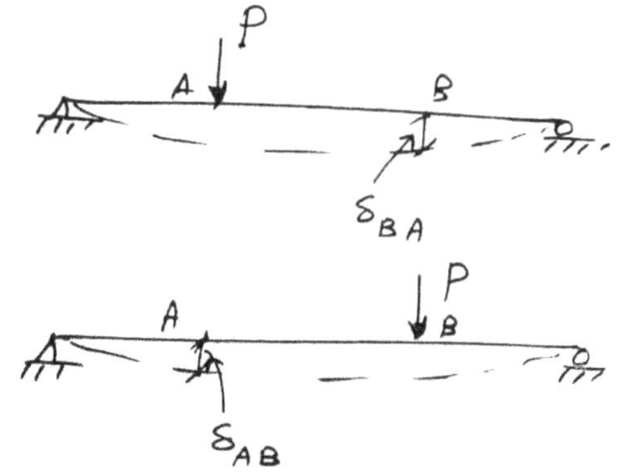

Betti's Law :

Betti's law states that

$$\frac{\delta_{BA}}{P_B} = \frac{\delta_{AB}}{P_A}$$

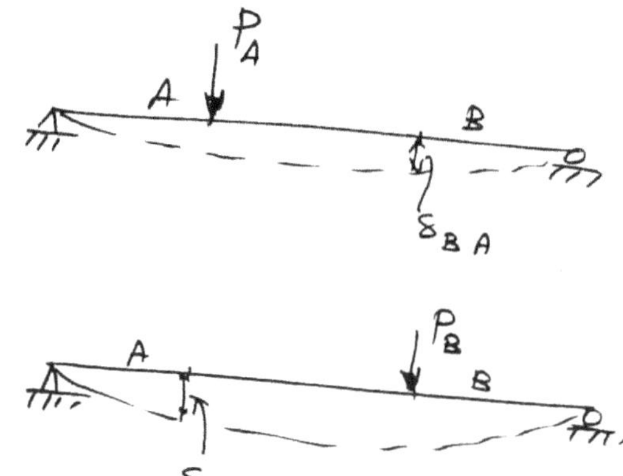

Notes:

1. Maxwell's law of reciprocal deflections is a special case of Betti's law when $P_A = P_B$.

2. James Clerk Maxwell first presented his law in 1864.

3. E. Betti generalized Maxwell's law in 1872.

4. The following two assumptions are used in the derivation of Maxwell's and Betti's laws:
   (a) the deformations are small.
   (b) the structure obeys Hooke's law (linear elasticity).

Types of Beams.

(1) Determinate Beams:

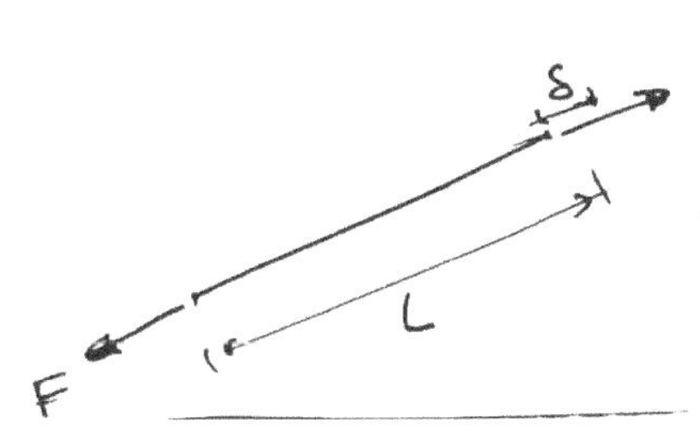

simply-supported beam
(simple beam)

overhanging beam

cantilever beam

(2) Indeterminate beams:

Propped Cantilever
with one span

Propped Cantilever
with two spans

Continuous Beam
with two spans

Continuous Beam
with three spans

Continuous Beam
with four spans

Elastic Supports:

① Cables:

L

δ

T

T

Deflection of Cable
(Extension)

$$\delta = \frac{TL}{EA}$$

② Springs

P

P

δ

$k \equiv$ spring stiffness

$$P = k\delta$$

$f \equiv$ spring flexibility , $f = \frac{1}{k}$

$\therefore P = \frac{1}{f}\delta \implies \boxed{\delta = fP}$

TRUSS MEMBERS:

F

F

δ

L

Deflection in a truss member

$$\delta = \frac{FL}{EA}$$

Degree of Indeterminacy :

Degree of indeterminacy = No. of unknowns — No. of equations

① Beams :

r : No. of reactions

$e = 3$: No. of equations of equilibrium

c = No. of hinges

No. of unknowns $= r$

No. of equations $= 3 + c$

The beam is indeterminate if $r > 3 + c$

Degree of indeterminacy $= r - (3 + c)$.

Example :

$r = 6$

$c = 1$

$3 + c = 4, \therefore 6 > 4$

$6 - 4 = \boxed{2}$.

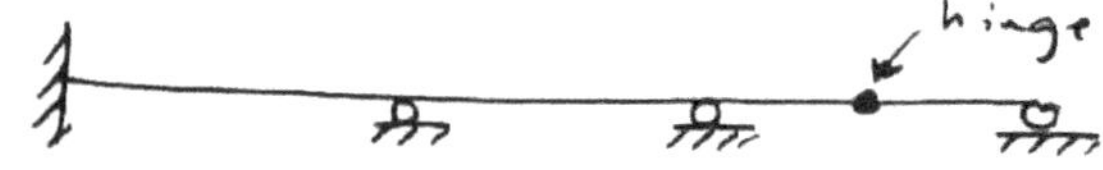

$\therefore$ The beam is indeterminate to the second degree.

② Trusses :

r : No. of reactions

j : No. of joints

m : No. of members

No. of unknowns $= r + m$

No. of equations $= 2j$

The truss is indeterminate if $r + m > 2j$.

Degree of indeterminacy $= (r + m) - 2j$.

Example :

$r = 3$

$j = 6$

$m = 10$

$r + m = 13 > 2j = 12$

$\therefore 13 - 12 = \boxed{1}$.

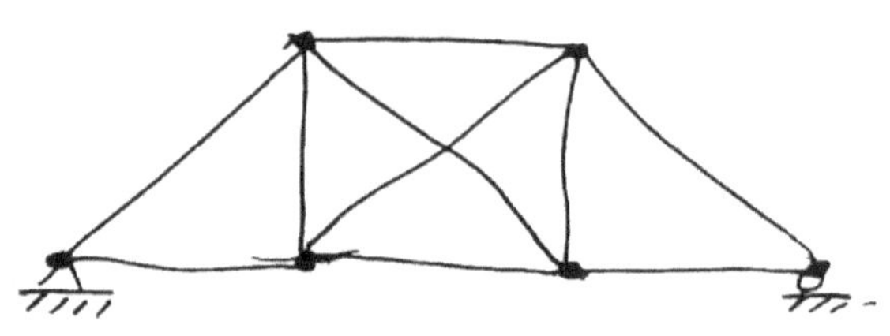

$\therefore$ The truss is indeterminate to the first degree.

③ __Frames__ : r : No. of reactions

j : No. of rigid joints

m : No. of members

c : No. of hinges

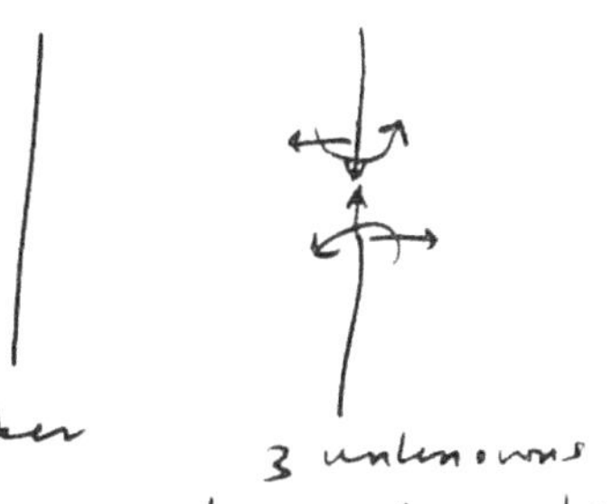

No. of unknowns $= r + 3m$

No. of equations $= 3j + c$

The frame (or beam) is indeterminate if $(r+3m) > (3j+c)$

Degree of indeterminacy $= (r+3m) - (3j+c)$

__Example__ : $r = 6$

$j = 6$

$m = 6$

$c = 0$

$r + 3m = 6 + 3(6) = 24$

$3j = 3(6) = 18$

$24 - 18 = 6$.

The frame is indeterminate to the __sixth degree__ .

__Another Method__ : (Transform) Separate the frame to two determinate frames. (and stable)

Degree of indeterminacy $= (3 \times$ No. of Cuts$)$ $-$ No. of hinges

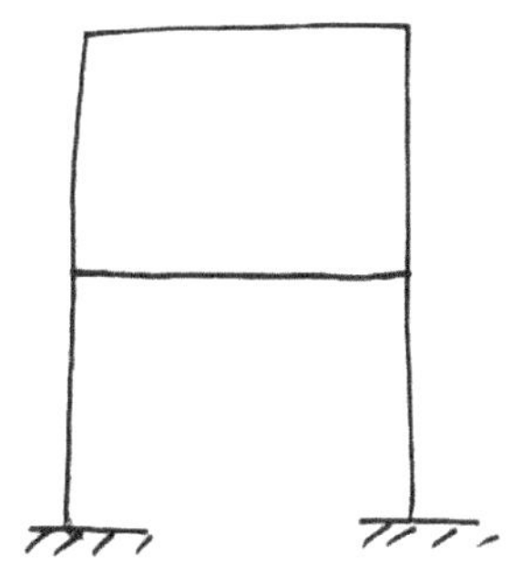

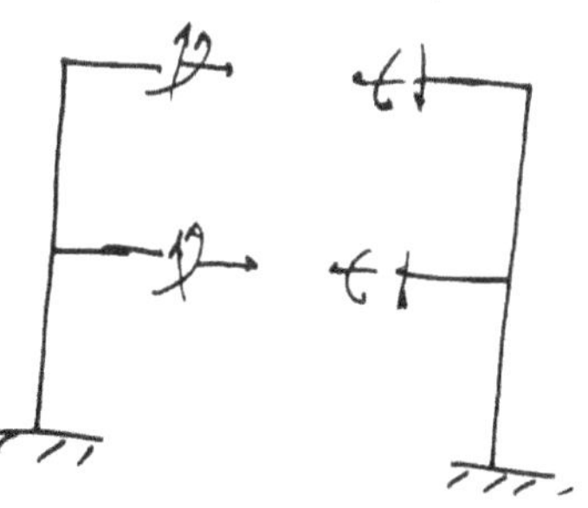

__Example__ : $r = 9$

$j = 9$

$m = 10$

$c = 0$

$r + 3m = 9 + 3(10) = 39$

$3j = 3(9) = 27$

$39 - 27 = 12$

∴ The frame is indeterminate to the __12th degree__ .

Example : $r = 9$
$j = 9$
$m = 10$
$c = 4$

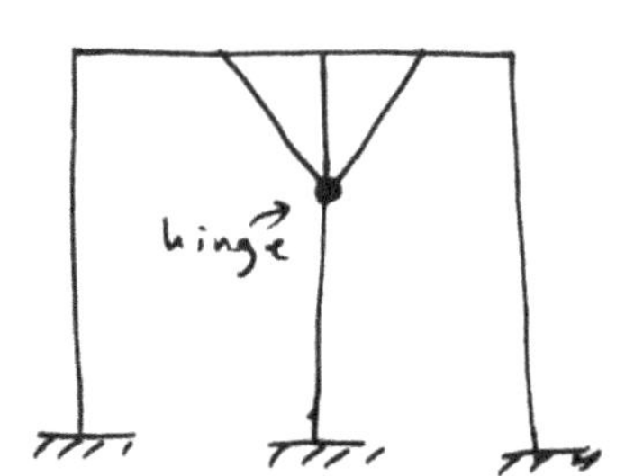

$r + 3m = 9 + 3(10) = 39$

$3j + c = 3(9) + 4 = 31$

$39 - 31 = 8$

The frame is indeterminate to the 8 th degree.

Example : $r = 9$
$j = 9$
$m = 10$
$c =$ No. of members meeting at
     the joint $- 1$
     $= 4 - 1 = 3$

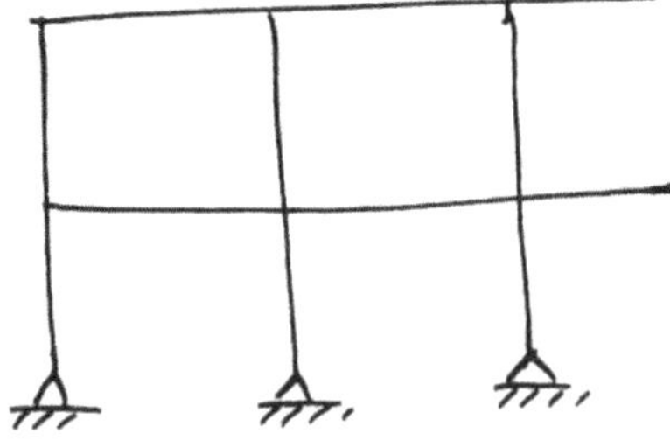

$r + 3m = 39$

$3j + c = 27 + 3 = 30$

$39 - 30 = 9$

The frame is indeterminate to the ninth degree.

Example :

The overhanging portions should
not be counted in the number
of members.

$r = 6$
$j = 9$
$m = 10$
$c = 0$
$r + 3m = 6 + 3(10) = 36$
$3j = 27$
$36 - 27 = 9$

The frame is indeterminate to the ninth degree.

Example:

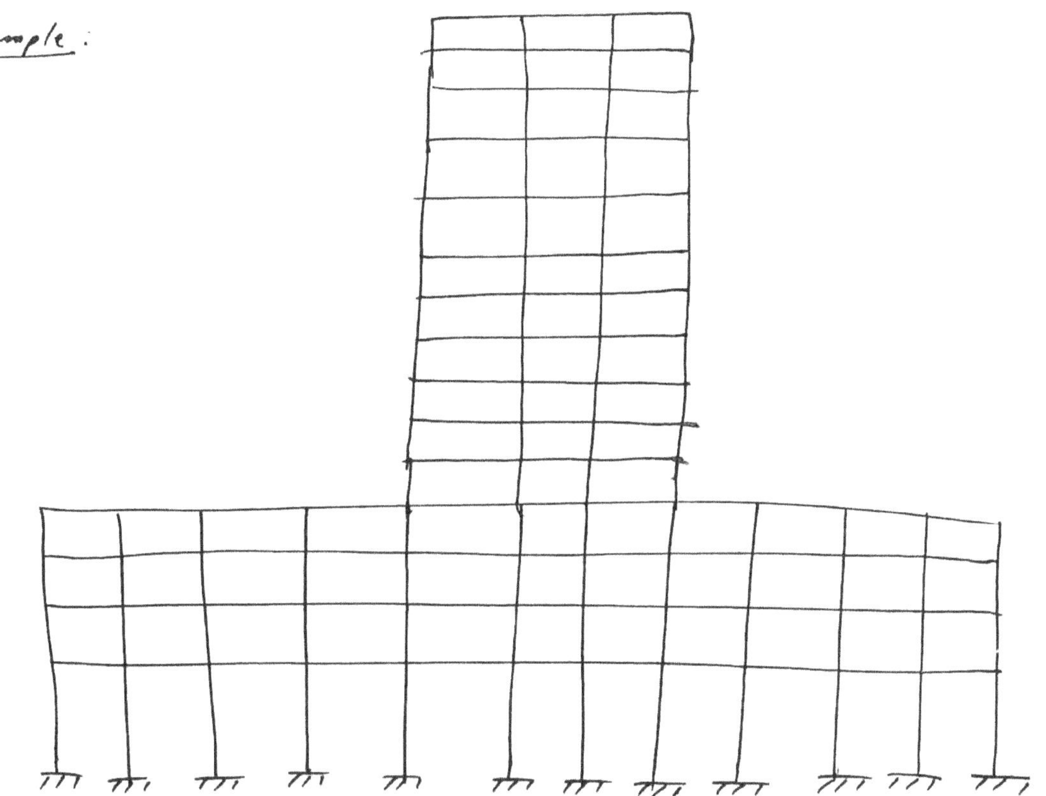

The frame can be separated into 12 stable and determinate parts by __77 cuts__ in the beams, it is indeterminate to the $3 \times 77 = 231\underline{st}$ degree.

No. of Cut $= 44 + 33 = $ __77 cuts__

Problems:

① 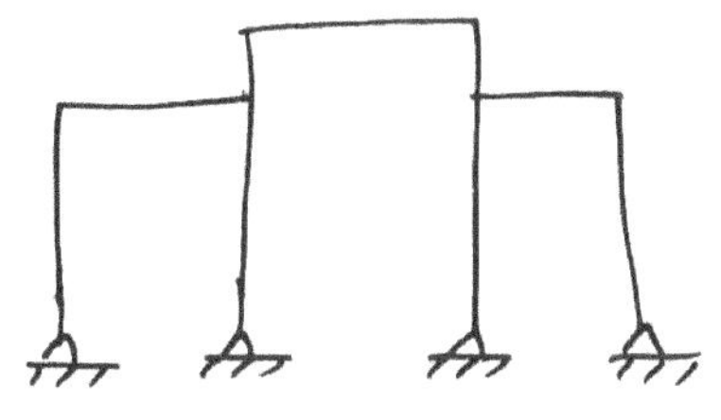

indeterminate to the $5\underline{th}$ degree.

② 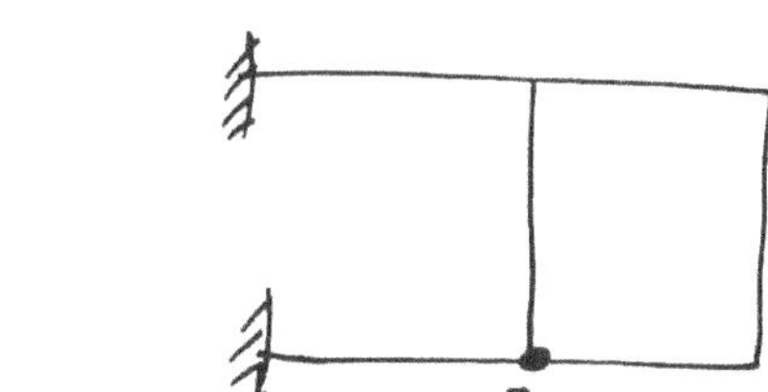

hinge

indeterminate to the $4\underline{th}$ degree

③ 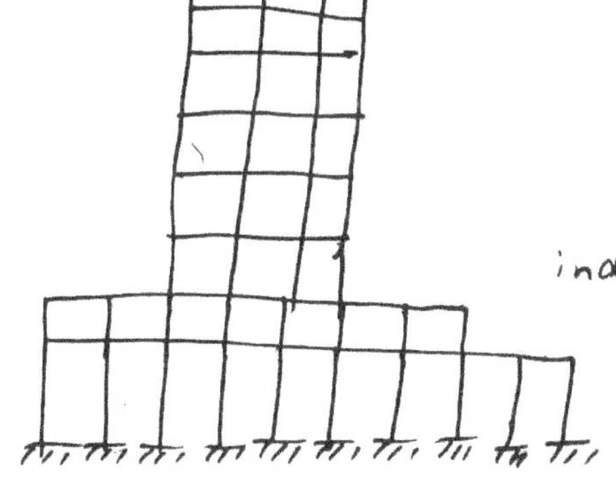

indeterminate to the $102\underline{nd}$ degree.

Approximate Analysis of Indeterminate Structures

① Introduction:

When a _model_ is used to represent any structure, then the analysis of it must satisfy **both** the conditions of equilibrium and compatibility of displacements at the joints. The compatibility conditions for a statically indeterminate structure can only be related to the loads, provided we know the material's modulus of elasticity and the size and shape of the members.

For design, however, we will **not** know a member's size, and so a statically indeterminate analysis cannot be considered. For analysis, a simpler model of the structure must be developed, one that is statically determinate. Once this model is specified, the analysis of it is called an _approximate analysis_.

By performing an approximate analysis, a preliminary design of the members of a structure can be made, and once this is complete, the more exact indeterminate analysis can then be performed and the design refined.

② Trusses:

Consider the truss shown in the figure. This is a truss with double diagonals. The points where the diagonal bars cross are **not** joints.

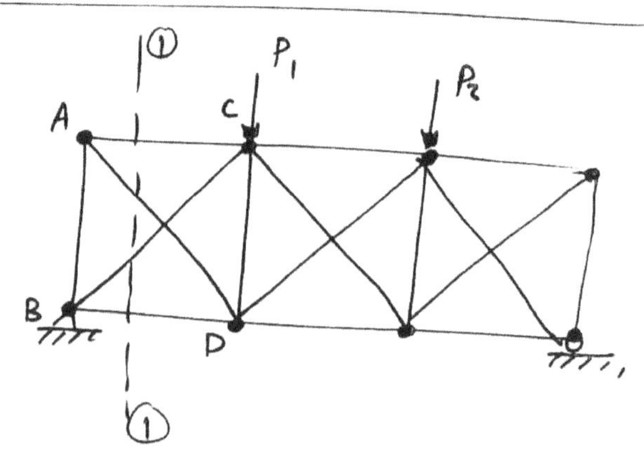

$$j = 8 \implies 2j = 16$$
$$\left.\begin{array}{l} r = 3 \\ m = 16 \end{array}\right\} \; m + r = 19$$

$\therefore$ Degree of indeterminacy $= 19 - 16 = 3$

$\therefore$ the truss has three internal redundants.

∴ We must make <u>three assumptions</u> in order to reduce the 1B2

truss to one that is statically determinate. These assumptions
will be made with regard to the cross-diagonals,
realizing that when one diagonal in a panel is in tension,
the corresponding cross-diagonal will be in compression. This
is evident from the figure below, when the "panel shear"
V is carried by the vertical component of tensile force
in member AD and the vertical component of compressive
force in member BC.

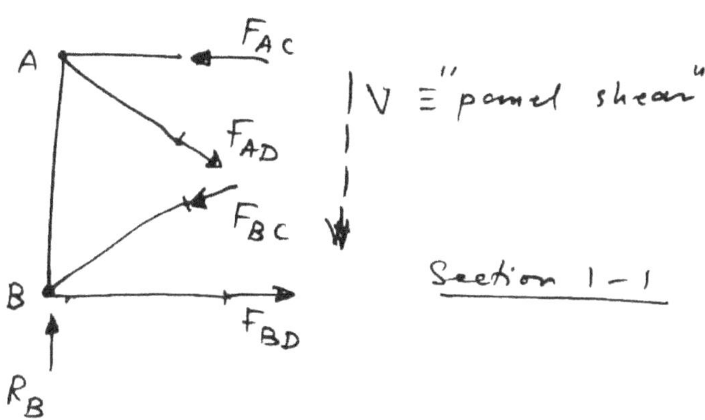

Two methods of analysis are generally used:

<u>Method 1</u>: If the diagonals are intentionally designed to be
<u>long</u> and <u>slender</u>, it is reasonable to assume that they
cannot support a compressive force; otherwise, they may
easily <u>buckle</u>. Hence, the panel shear is resisted entirely
by the tension diagonal, whereas the compressive diagonal
is assumed to be a zero-force member.

   i.e. <u>assume</u> $\boxed{F_{BC} = 0}$.

<u>Method 2</u>: If the diagonal members are intended to be
constructed from large rolled sections, they may be equally
capable of supporting a tensile and compressive force.
Hence, we can assume that the tension and compression
diagonals each carry half the panel shear.

   i.e. <u>assume</u> $\boxed{F_{BC} = F_{AD} = F}$

③ Vertical Loads on Building Frames:

Building frames often consist of girders that are <u>rigidly</u> connected to columns so that the entire structure is better able to resist the effects of lateral forces due to wind and earthquake. An example of such a rigid framework, often called a <u>building bent</u>, is shown in the figure.

In this section, we will establish a method for analyzing (approximately) the forces in building frames due to <u>vertical loads</u>. the simplifying assumptions made to reduce a frame from a statically indeterminate structure to one that is statically determinate are based on the way the structure deforms under the load.

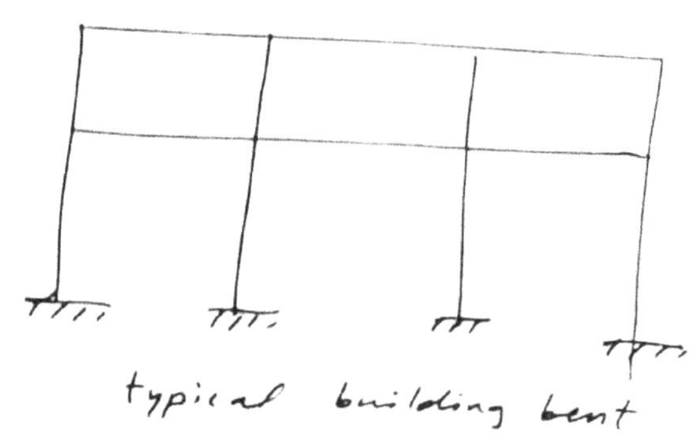

typical building bent

(a) Consider the fixed beam shown in the figure and subjected to a vertical load.

The beam is indeterminate to the third degree. therefore, three assumptions must be made for an approximate analysis of the beam.

From the shown deflection curve, we observe two inflection points at 0.21ℓ from each end. the bending moment is zero at these two points.

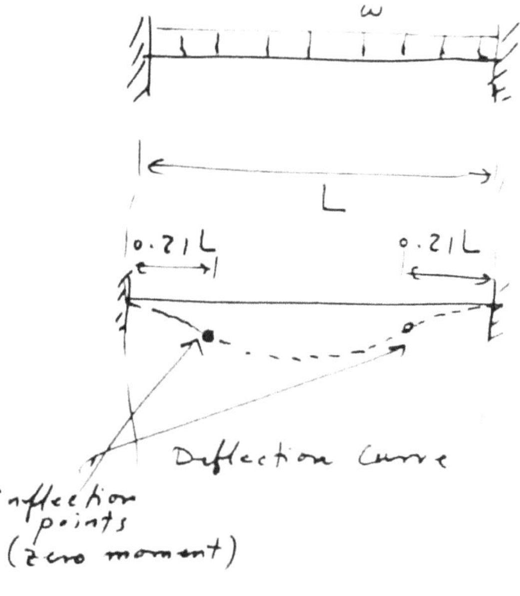

(b) Consider the simple beam shown in the figure and subjected to the vertical load shown. In this case, the points of zero moment occur at the supports.

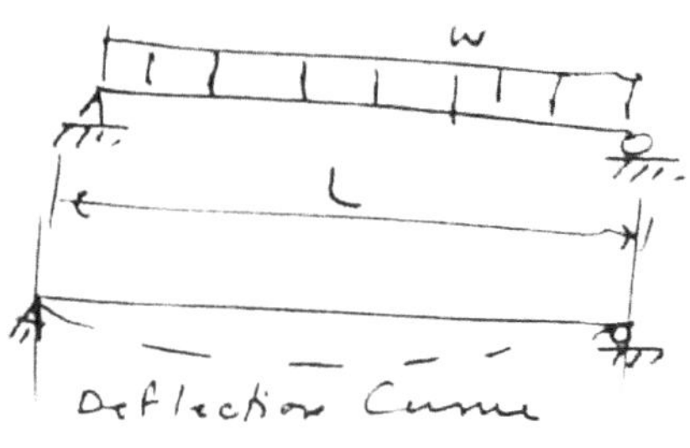

(c) For a real beam, the supports will be flexible (i.e. they will neither be simple or fixed supports), and therefore the points of zero moment are assumed to occur at the average point between the two extremes, i.e. at

$$\frac{0.21L + 0}{2} \cong 0.1L$$ from each support. Furthermore, an exact analysis of frames supporting vertical loads indicates that the axial forces in the beam are negligible.

Therefore, the following three assumptions are used for the approximate analysis of beams subjected to vertical loads.

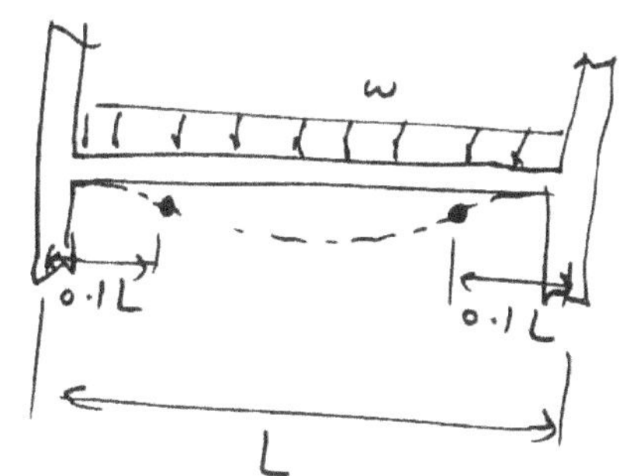

1. There is a hinge (zero moment), at $0.1L$ from the left support.

2. There is a hinge (zero moment), at $0.1L$ from the right support.

3. There is no axial force in the beam.
   The above three assumptions will reduce the beam to one that is statically determinate.

④ Lateral Loads on Building Frames: the Portal Method:

Consider the building bent shown in the figure, along with its deflection curve.
Inflection points occur at the locations shown in the figure.
Therefore, for an approximate analysis of the frame, we can assume hinges to be present at the centers of beams and columns.

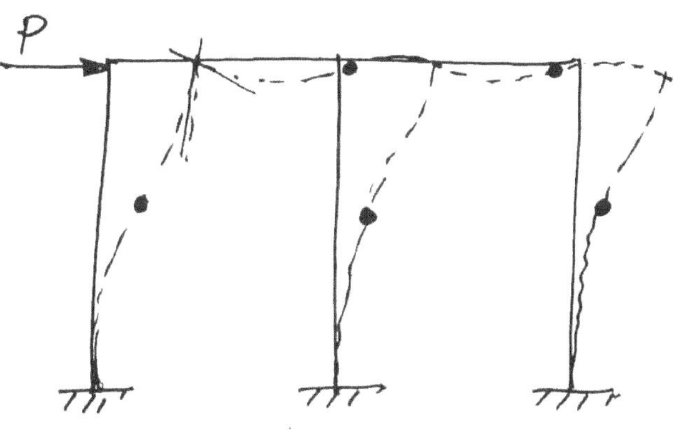

However, for the example given, the frame is indeterminate to the sixth degree, and we have made only five assumption (i.e. five hinges are assumed). We need a sixth assumption to reduce the frame to one that is statically determinate.

According to the _portal method_, the interior columns carry _twice_ the shear V as the _exterior_ columns (at the hinges).

To summarize, an approximate analysis of building frames subjected to lateral loads using the portal method, employs the following assumptions:

1. A hinge is placed at the center of each beam.
2. A hinge is placed at the center of each column.
3. At a given floor level, the shear at the interior column hinge is twice that at the exterior column hinge (since the frame is considered to be a superposition of portal frames).

For the example given, the third assumption is shown schematically in the figure below:

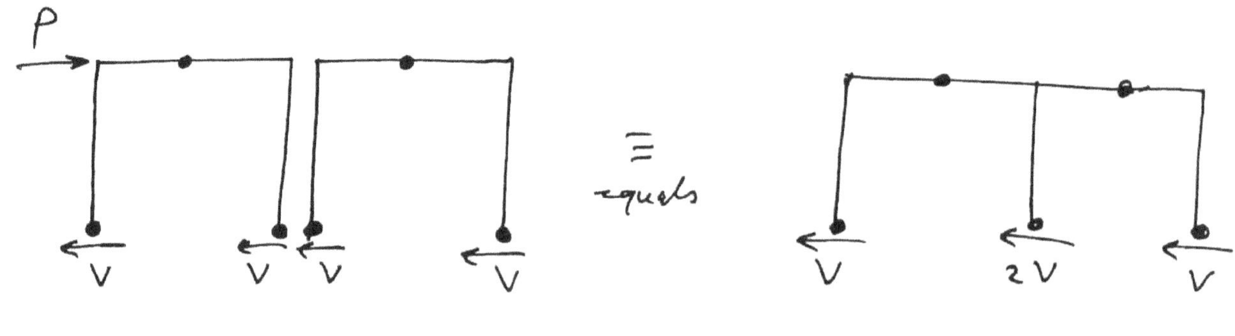

⑤ Lateral Loads on Building Frames — The Cantilever Method:

The cantilever method assumes that the axial stress in a column is proportional to its distance from the centroid of all the column areas at a given floor level. See the figure below.

To counteract the possible tipping, the axial forces (or stress) in the columns will be tensile on one side of the centroidal axis and compressive on the other side, as shown in the figure.

Therefore, the cantilever method is appropriate if the frame is tall and slender, or has columns with different cross-sectional areas.

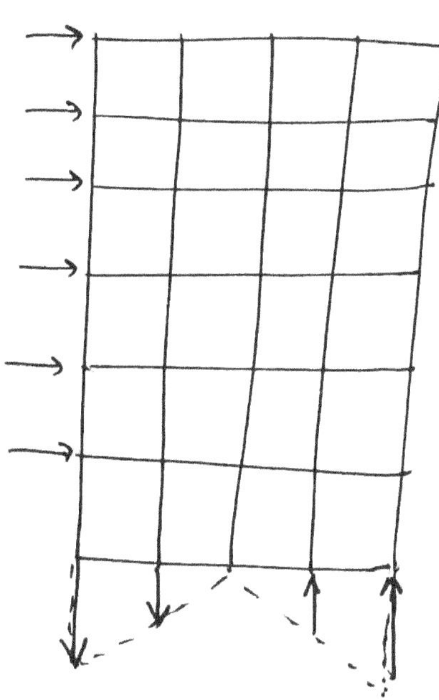

Building Frame

To summarize, an approximate analysis of building frames subjected to lateral loads using the cantilever method, employs the following assumptions:

1. A hinge is placed at the center of each beam.

2. A hinge is placed at the center of each column.

3. The axial stress in a column is proportional to its distance from the centroid of the cross-sectional areas of the columns at a given floor level. Since stress equals force per area, then in the special case of the columns having equal cross-sectional areas, the force in a column is also proportional to its distance from the centroid of the column areas.

(6) Examples:

Example – 1 – :

Determine (approximately) the reactions at the base of the columns of the frame shown in the figure. Use the portal method of analysis.

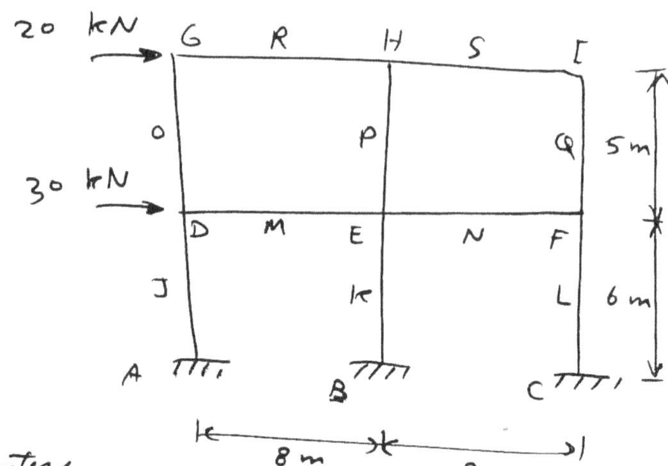

Solution :

First, place **hinges** at the centers of the beams and columns, i.e.

hinges are placed at the ten locations

$$J, k, L, O, P, Q, M, N, R, S.$$

the frame is indeterminate to the 12th degree.

∴ We need **two** more assumptions.

(1) Apply assumption (3) to the top floor level, at O, P, Q :

$$\xrightarrow{+} \Sigma F_x = 0:$$

$$20 - V - 2V - V = 0$$

$$\Rightarrow V = 5 \ kN.$$

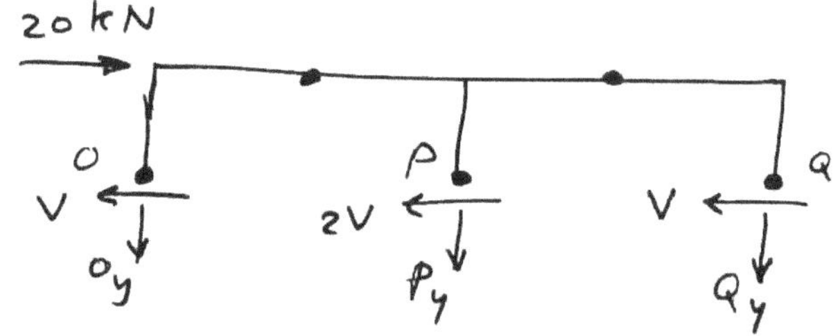

(2) Apply assumption (3) to the lower floor level, at J, k, L.

$$\xrightarrow{+} \Sigma F_x = 0:$$

$$20 + 30 - V' - 2V' - V' = 0$$

$$\Rightarrow V' = 12.5 \ kN.$$

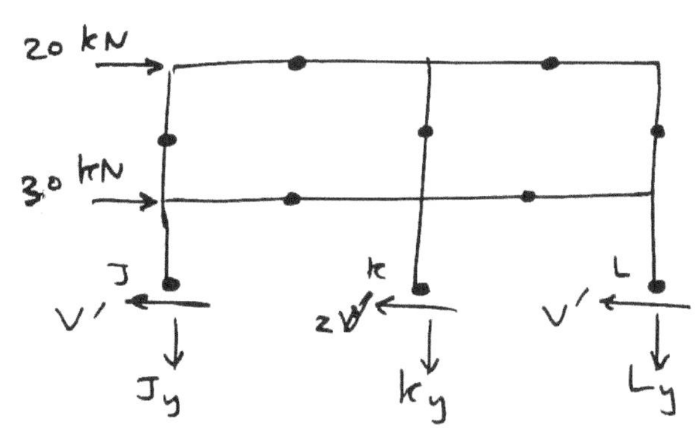

Now, we can proceed with a determinate analysis of the frame.
Start at the corners where the loads are applied.

<u>Section OGR</u> :

$\xrightarrow{+} \; \Sigma F_x = 0$:

$\quad -R_x - 5 + 20 = 0$

$\qquad R_x = 15 \; kN$.

$+\circlearrowleft \; \Sigma m_o = 0$:

$\quad -20 \left(\frac{5}{2}\right) + (15)\left(\frac{5}{2}\right) + R_y (4) = 0$

$\qquad R_y = 3.125 \; kN$.

$+\uparrow \; \Sigma F_y = 0$: $\quad -O_y + 3.125 = 0 \implies O_y = 3.125 \; kN$.

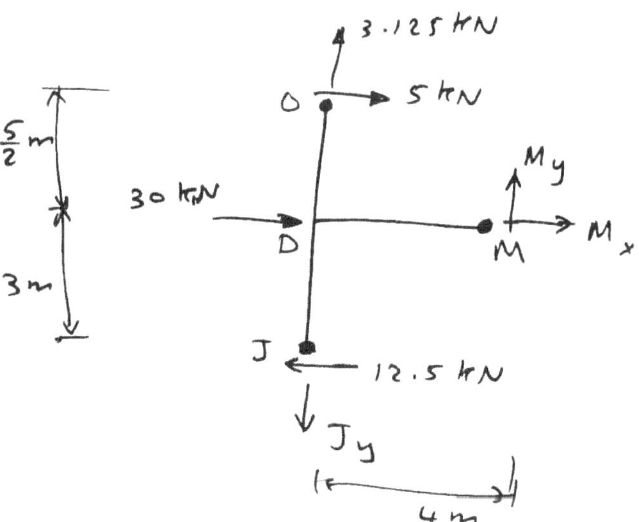

<u>Section OMJ</u> :

$\xrightarrow{+} \; \Sigma F_x = 0$:

$\quad 5 + M_x - 12.5 + 30 = 0$

$\qquad M_x = -22.5 \; kN$.

$+\circlearrowleft \; \Sigma m_J = 0$:

$\quad -30(3) - 5(5.5) - (-22.5)(3)$

$\qquad + M_y (4) = 0$

$\quad \therefore \; M_y = 12.5 \; kN$.

$+\uparrow \; \Sigma F_y = 0$: $\quad -J_y + 12.5 + 3.125 = 0$

$\qquad J_y = 15.625 \; kN$.

<u>Section JA</u> :

$\xrightarrow{+} \; \Sigma F_x = 0$: $\boxed{A_x = 12.5 \; kN}$.

$+\uparrow \; \Sigma F_y = 0$: $\boxed{A_y = 15.625 \; kN}$

$+\circlearrowleft \; \Sigma m_A = 0$: $\quad M_A - 12.5(3) = 0$

$\qquad \boxed{M_A = 37.5 \; kN \cdot m}$.

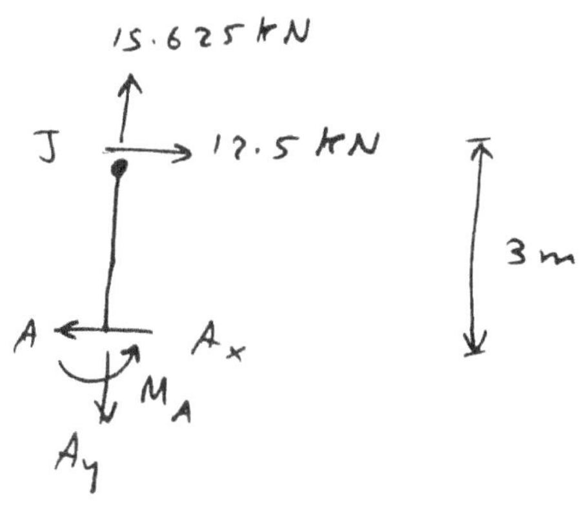

Section RPS:

$\xrightarrow{+}\ \Sigma F_x = 0:$

$15 - 10 - S_x = 0$

$S_x = 5\ kN.$

$+\circlearrowleft\ \Sigma M_P = 0:$

$3.125(4) - 15\left(\dfrac{5}{2}\right) + (5)\left(\dfrac{5}{2}\right) + S_y(4) = 0$

$\therefore\ S_y = 3.125\ kN.$

$+\uparrow\ \Sigma F_y = 0:\quad -3.125 + P_y + 3.125 = 0$

$P_y = 0.$

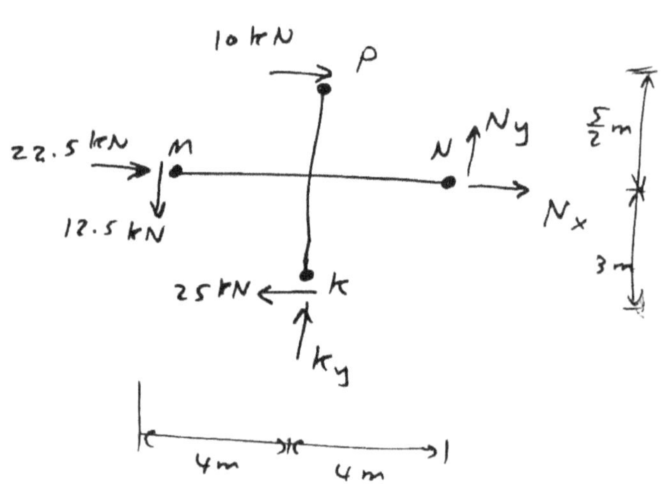

Section MPNK:

$\xrightarrow{+}\ \Sigma F_x = 0:$

$22.5 - 25 + 10 + N_x = 0$

$N_x = -7.5\ kN.$

$+\circlearrowleft\ \Sigma M_K = 0:$

$12.5(4) - 22.5(3) - 10(5.5)$

$\qquad -(-7.5)(3) + N_y(4) = 0$

$\therefore\ N_y = 12.5\ kN.$

$+\uparrow\ \Sigma F_y = 0:\quad -12.5 + k_y + 12.5 = 0$

$k_y = 0.$

Section KB:

$\xrightarrow{+}\ \Sigma F_x = 0:\quad \boxed{B_x = 25\ kN}$

$+\uparrow\ \Sigma F_y = 0:\quad \boxed{B_y = 0}$

$+\circlearrowleft\ \Sigma M_B = 0:\quad M_B - 25(3) = 0$

$$\boxed{M_B = 75\ kN\cdot m}$$

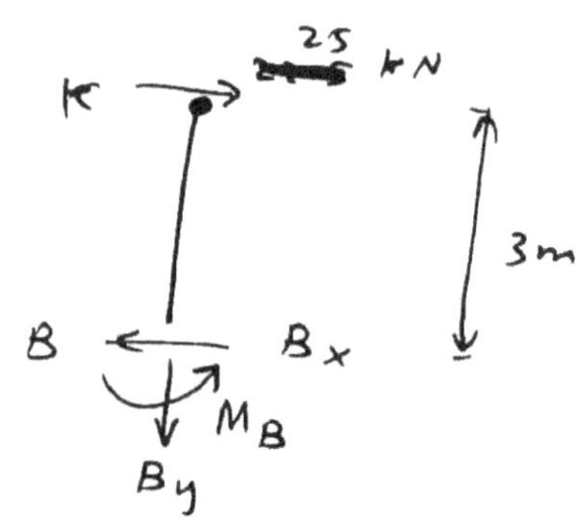

Section SIQ:

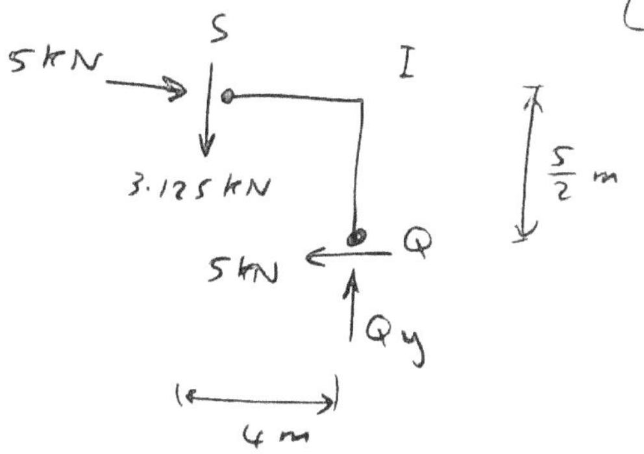

$+\uparrow \Sigma F_y = 0:$ $Q_y = 3.125 \ kN$.

check that $M_Q = 0:$

$+\circlearrowleft \Sigma M_Q = 0:$ $3.125(4) - 5\left(\frac{5}{2}\right) = 0$

$$0 = 0 \quad \smile$$

Section QNL:

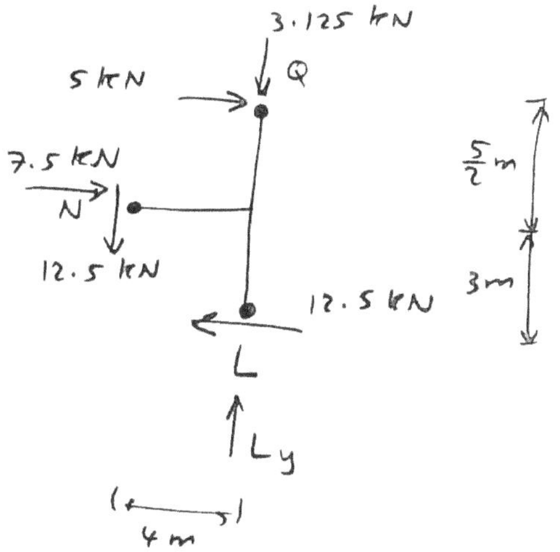

$+\uparrow \Sigma F_y = 0:$

$\quad L_y - 12.5 - 3.125 = 0$

$\quad \therefore \ L_y = 15.625 \ kN$.

check that $M_L = 0:$

$+\circlearrowleft \Sigma M_L = 0:$

$\quad 12.5(4) - 7.5(3) - 5(5.5) = 0$

$$0 = 0 \quad \smile$$

Section LC:

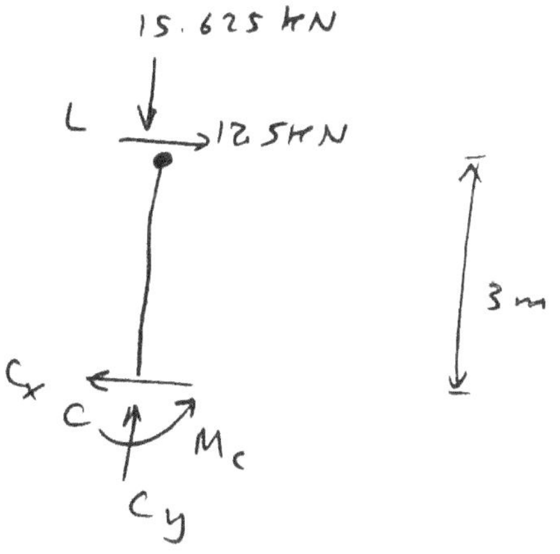

$\xrightarrow{+} \Sigma F_x = 0:$ $\boxed{C_x = 12.5 \ kN}$.

$+\uparrow \Sigma F_y = 0:$ $\boxed{C_y = 15.625 \ kN}$.

$+\circlearrowleft \Sigma M_C = 0:$ $M_C - 12.5(3) = 0$

$$\boxed{M_C = 37.5 \ kN \cdot m}$$

the reactions are shown on the frame below:

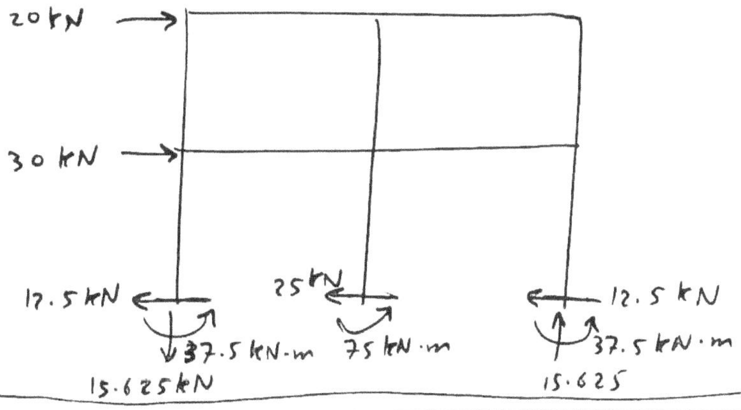

<u>Example - 2 -</u> :

Repeat Example -1- using the cantilever method. Compare the results of both methods.

<u>Solution</u> :

Since the columns have <u>equal</u> cross-sectional areas, we can use the forces in the columns (instead of the stresses).

First, locate the centroid of the columns' cross-sectional areas.

let $A \equiv$ area of each column

$$\bar{x} = \frac{\Sigma \tilde{x} A}{\Sigma A} = \frac{0(A) + 8(A) + 16(A)}{A + A + A}$$

$$= \frac{0 + 8 + 16}{3}$$

$$= 8 \text{ m}.$$

(this is clear <u>only</u> in this example).

Second, place <u>hinges</u> at the center of each beam and column.

Like the previous example, we need <u>two</u> more assumptions.

(P) Assume that the axial force in each column is proportional to its distance from the centroid at level OPQ.

It is clear from the figure that $P_y = 0$ (from symmetry).

From similar triangles,

$$\frac{Q_y}{8} = \frac{O_y}{8}$$

$$\Rightarrow \boxed{Q_y = O_y} \quad \text{———} \quad (1)$$

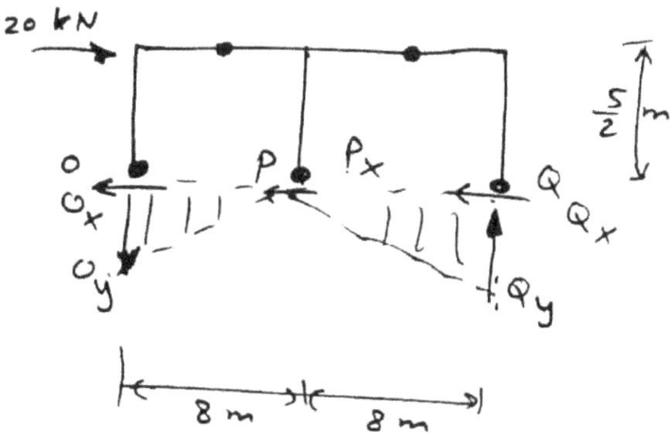

$+\circlearrowleft \Sigma M_o = 0 : \quad Q_y(16) - 20\left(\frac{5}{2}\right) = 0$

$$Q_y = 3.125 \text{ kN}.$$

$$\therefore O_y = 3.125 \text{ kN}.$$

2) Assume that the axial force in each column is proportional to its distance from the centroidal axis at level JKL.

It is clear from symmetry that $\underline{K_y = 0}$.

From similar triangles:

$$\frac{L_y}{8} = \frac{J_y}{8}$$

$$\implies \boxed{L_y = J_y}$$

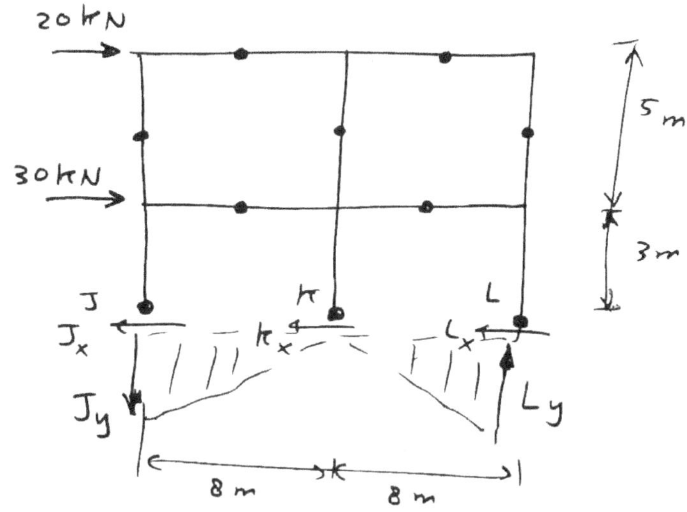

$\circlearrowright \Sigma m_J = 0$:

$$L_y (16) - 20 (8) - 30(3) = 0$$
$$L_y = 15.625 \ kN$$
$$\therefore \ J_y = 15.625 \ kN$$

Next, we proceed with a determinate analysis of the frame. Start at the corners where the loads are applied.

<u>Section OGR</u>:

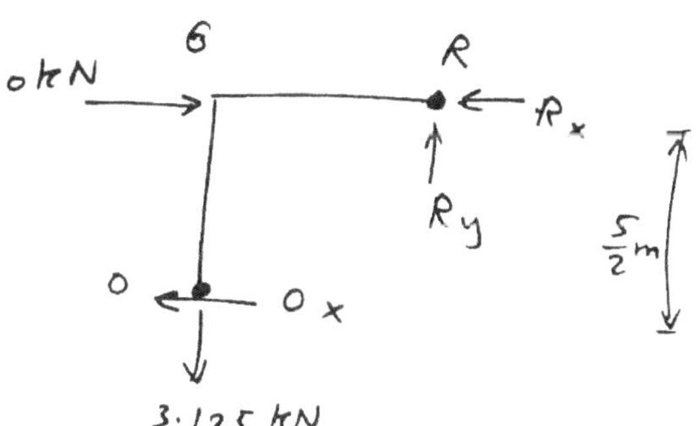

$+\uparrow \Sigma F_y = 0$: $\quad R_y = 3.125 \ kN$.

$+\circlearrowleft \Sigma m_o = 0$:

$$3.125 (4) + R_x \left(\frac{5}{2}\right) - 20 \left(\frac{5}{2}\right) = 0$$
$$\therefore \ R_x = 15 \ kN.$$

$\xrightarrow{+} \Sigma F_x = 0$: $\quad -O_x - 15 + 20 = 0$

$$O_x = 5 \ kN.$$

<u>Section OMJ</u>:

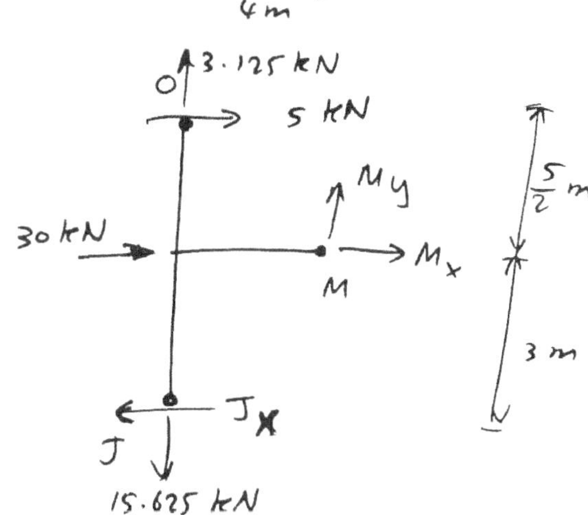

$+\uparrow \Sigma F_y = 0$:

$$3.125 + M_y - 15.625 = 0$$
$$M_y = 12.5 \quad kN.$$

$+\circlearrowleft \Sigma m_J = 0$:

$$-30(3) - 5(5.5) - M_x(3) + 12.5 (4) = 0$$
$$M_x = -22.5 \quad kN.$$

$\xrightarrow{+} \Sigma F_x = 0: \quad -J_x + (-22.5) + 5 + 30 = 0$

$$J_x = 12.5 \ kN$$

Section JA:

$\xrightarrow{+} \Sigma F_x = 0: \boxed{A_x = 12.5 \ kN}$

$+\uparrow \Sigma F_y = 0: \ A_y = \boxed{15.625 \ kN}$

$+\circlearrowleft \Sigma M_A = 0: \quad M_A - \dfrac{12.5}{}(3) = 0$

$$\boxed{M_A = 37.5 \ kN \cdot m}$$

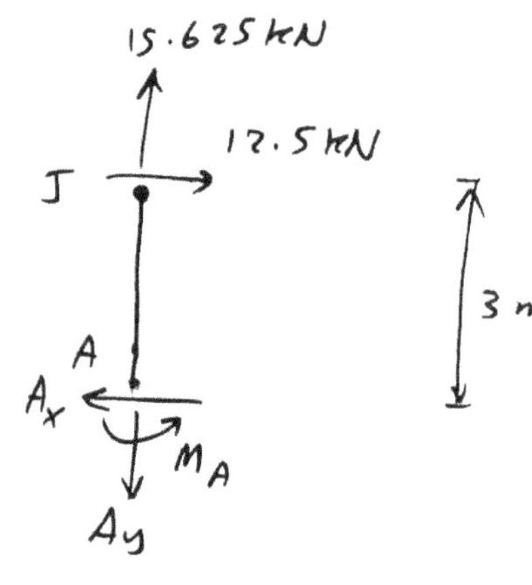

Section RPS:

$+\uparrow \Sigma F_y = 0: \ S_y = 3.125 \ kN$

$+\circlearrowleft \Sigma M_p = 0:$

$3.125(4) - S_x\left(\dfrac{5}{2}\right) - 15\left(\dfrac{5}{2}\right) + 3.125(4) = 0$

$$S_x = -5 \ kN$$

$\xrightarrow{+} \Sigma F_x = 0: \quad 15 - P_x + (-5) = 0$

$$P_x = 10 \ kN$$

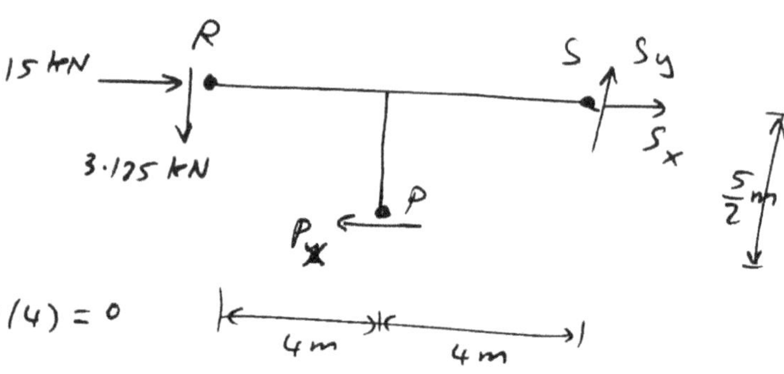

Section MPNK:

$+\uparrow \Sigma F_y = 0: \ N_y = 12.5 \ kN$

$\circlearrowleft \Sigma M_k = 0:$

$22.5(4) - N_x(3) - 10(5.5)$

$\quad - 22.5(3) + 12.5(4) = 0$

$$\therefore \ N_x = -7.5 \ kN$$

$\xrightarrow{+} \Sigma F_x = 0: \ 22.5 - k_x + (-7.5) = 0$

$$k_x = 15 \ kN$$

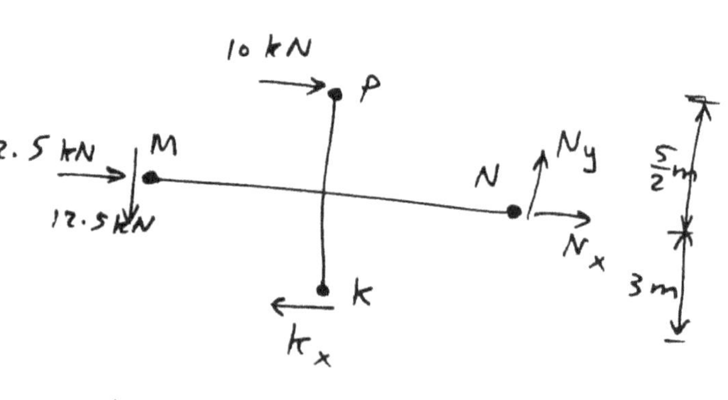

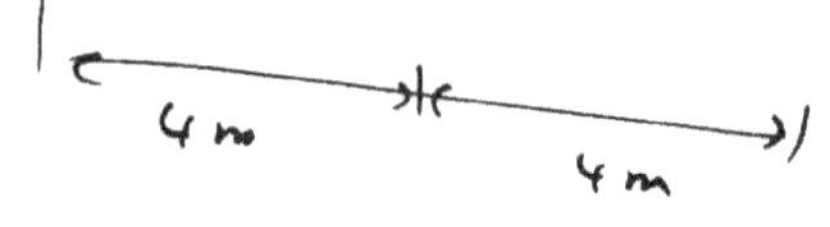

Section KB:

$\xrightarrow{+} \Sigma F_x = 0:$ $\boxed{B_x = 15 \text{ kN}}$

$+\uparrow \Sigma F_y = 0:$ $\boxed{B_y = 0}$

$+\circlearrowleft \Sigma m_B = 0:$ $M_B - 15(3) = 0$

$$\boxed{M_B = 45 \text{ kN} \cdot \text{m}}$$

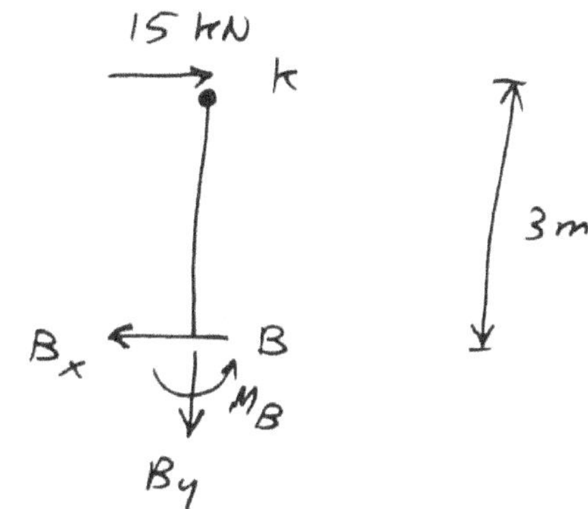

Section SIQ:

$\xrightarrow{+} \Sigma F_x = 0:$ $Q_x = 5 \text{ kN}$

Check that $M_Q = 0$:

$+\circlearrowleft \Sigma m_Q = 0:$

$$3.125(4) - 5\left(\frac{5}{2}\right) = 0$$

$$0 = 0 \;\smile$$

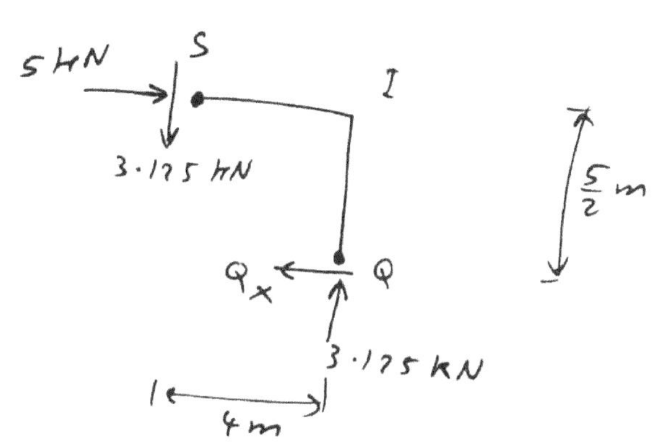

Section QNL:

$\xrightarrow{+} \Sigma F_x = 0:$ $-L_x + 7.5 + 5 = 0$

$$L_x = 12.5 \text{ kN}$$

$+\circlearrowleft \Sigma m_L = 0:$ (Check):

$$-5(5.5) - 7.5(3) + 12.5(4) = 0$$

$$0 = 0 \;\smile$$

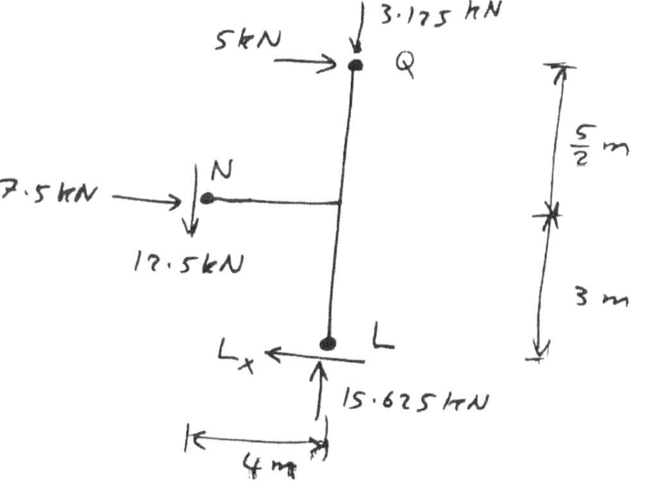

Section LC:

$\xrightarrow{+} \Sigma F_x = 0:$ $\boxed{C_x = 12.5 \text{ kN}}$

$+\uparrow \Sigma F_y = 0:$ $\boxed{C_y = 15.625 \text{ kN}}$

$+\circlearrowleft \Sigma m_c = 0:$ $M_c - 12.5(3) = 0$

$$\boxed{M_c = 37.5 \text{ kN} \cdot \text{m}}$$

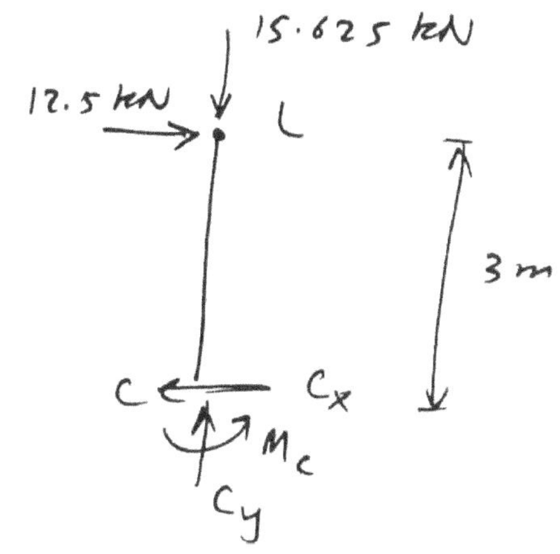

the reactions are shown on the frame below.

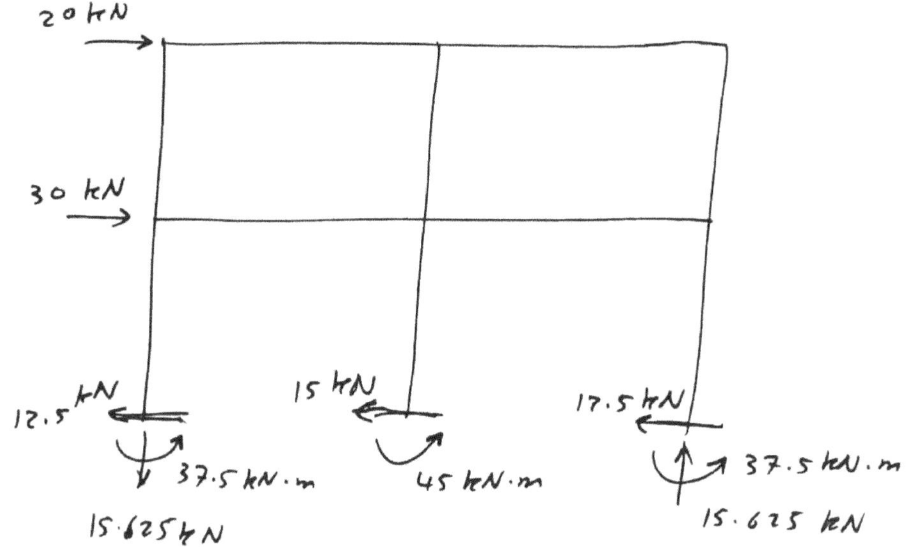

Remarks: the frame was analyzed using a computer program. the following table shows a comparison of the results:

Reaction	Portal Method	Cantilever Method	Computer Solutions
A_x	−12.5	−12.5	−16.0
A_y	−15.625	−15.625	−13.9
M_A	+37.5	+37.5	+57.0
B_x	−25.0	−15.0	−18.8
B_y	0	0	0
M_B	+75.0	+45.0	+63.4
C_x	−12.5	−12.5	−16.0
C_y	+15.625	+15.625	+13.9
M_C	+37.5	+37.5	+57.0

<u>Example – 3 –</u> :

Determine (approximately) the moments at the joints E and C caused by members EF and CD of the building bent shown in the figure.

<u>Solution</u> :

We assume <u>hinges</u> at $0.1L$ from each support (for the two beams) and neglect axial forces in the beam.

$$0.1 L = 0.1 (20) = 2 \text{ ft.}$$

Consider the top beam (between the hinges):

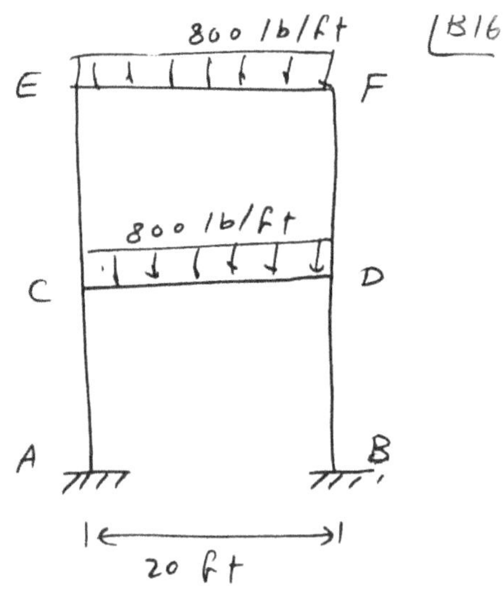

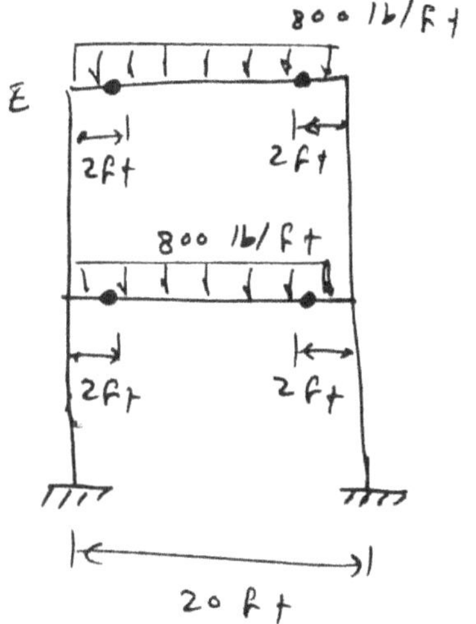

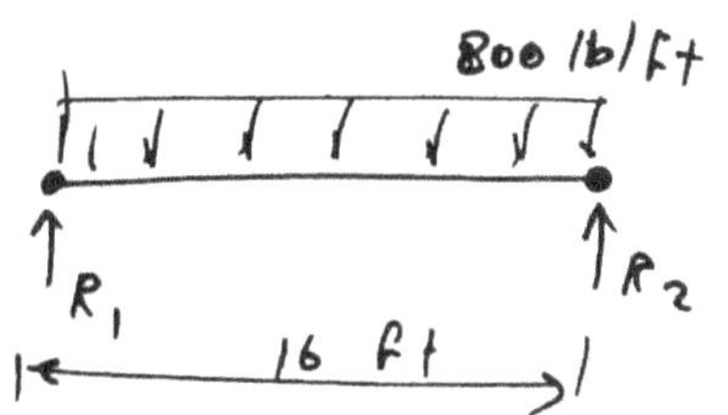

Axial forces are neglected (in this approximate analysis).

By symmetry, $R_1 = R_2 = \dfrac{800(16)}{2} = 6400$ lb.

Consider the section between joint E and the hinge:

$+\circlearrowleft \ \Sigma M_E = 0$:

$$M_E - 800(2)(1) - 6400(2) = 0$$

$$M_E = 14400 \quad \text{lb. ft}$$

$$M_E = 14.4 \ \text{k–ft} .$$

Also $M_C = 14.4$ k–ft (similarly).

Chapter 11: Method of Consistent Deformations
(The General Method)

11.1 Introduction: (The Force Method)

until pages (234-235)

* In addition to the equilibrium equations, we employ deformation (deflection) equations. The deformations must be consistent. i.e (no deflections at supports and no rotations at fixed ends)

Methods of analyzing indeterminate structures can be categorized as follows.

   ① Force Methods : The unknowns are forces.
      - Method of Consistent Deformations.
      - Three-Moment Equation
      - Method of Least Work (Castigliano's theorems)
      - Column Analogy
      - Flexibility Matrix Method

   ② Deformation Methods : (Displacement Methods) The unknowns are deflections.
      - Slope-Deflection Method
      - Moment Distribution Method.
      - Stiffness Matrix Method.

History:
1864: James Clerk Maxwell.
1874: Otto Mohr
The Maxwell–Mohr Method
"Method of Redundant

Method of Consistent Deformations

11.2 Basic Principles :

Example :

(1) Determine the degree of indeterminacy
   (i.e. number of redundants).

No. of equations = 3 (3 equations of equilibrium).

No. of unknowns $r = 4$

Degree of indeterminacy = $4 - 3 = 1$ (one redundant).

Using $\Sigma F_x = 0$: $\implies R_{A_x} = 0$

$\implies$ No. of equations = 2 (2 remaining equations of equilibrium)
   No. of unknowns = 3

Degree of indeterminacy = $3 - 2 = 1$ (still, one redundant)

$\therefore$ beam is indeterminate to the first degree.

Redundants are restraints that can be removed without impairing the load-supporting capacity of the structure.

No. of redundants = Degree of indeterminacy.

* Remove enough restraints (reactions) to obtain a stable and determinate structure.

① removing R_B

② Removing M_A

③ adding a hinge

Solution :

Consider Case ① : We remove R_B. the determinate and stable structure thus obtained is called the primary structure.

(Use the principle of superposition).

actual structure = primary structure + second structure (R_B)

$$\boxed{\delta_B = 0}$$

$$\boxed{\delta_{BC} + R_B \, \delta_{BB} = 0}$$ Equation of consistent deformations

(Compatibility Equation).

∴ Calculate δ_{BC} and $\boxed{\delta_{BB}}$ and substitute in the above equation, then solve for R_B.

flexibility coefficient

① Primary Structure . Calculate δ_{BC}.

Use the moment-area theorems.

$$\delta_{BC} = -\frac{1}{2}\left(\frac{PL}{2EI}\right)\left(\frac{L}{2}\right)\left(\frac{L}{2} + \frac{2}{3}\left(\frac{L}{2}\right)\right)$$

$$= -\frac{PL^2}{8EI} \cdot \frac{5L}{6} = -\frac{5PL^3}{48EI}$$

② <u>Second Structure:</u>

$$\delta_{BB} = +\frac{1}{2}\left(\frac{L}{EI}\right) \cdot L \cdot \frac{2}{3}L$$

$$= \frac{L^3}{3EI}$$

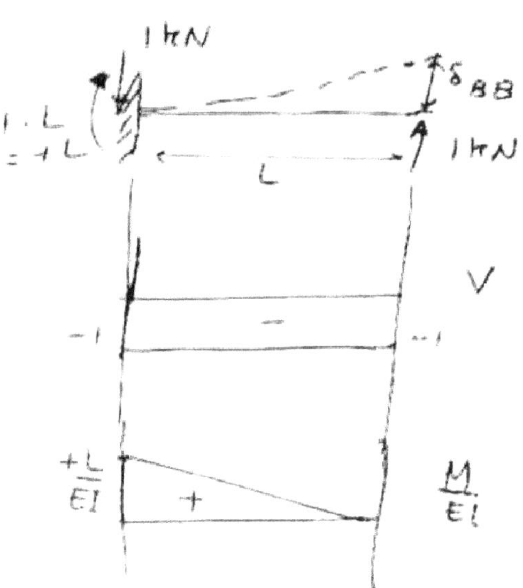

③ Substitute in the equation of consistent deformations :

$$\delta_{BC} + R_B\,\delta_{BB} = 0$$

$$-\frac{5PL^3}{48EI} + R_B\left(\frac{L^3}{3EI}\right) = 0$$

$$R_B = \frac{5PL^3}{48EI} \cdot \frac{3EI}{L^3} = \frac{5P}{16} = \boxed{\frac{5}{16}P = R_B}\ \text{the redundant.}$$

④ Use the equations of equilibrium to find the other reactions.

$+\uparrow\ \Sigma F_y = 0$:

$$R_A + \frac{5}{16}P - P = 0$$

$$R_A = \frac{11}{16}P\uparrow$$

$+\circlearrowright\ \Sigma M_A = 0:\quad M_A - P\frac{L}{2} + \frac{5P}{16}L = 0$

$$M_A = \frac{1}{2}PL - \frac{5}{16}PL$$

$$\implies M_A = \frac{3PL}{16}$$

<u>Steps in the Method of Consistent Deformations :</u>

(1) Determine the number of redundants that the structure possess.

(2) Remove enough redundants to form a <u>stable</u> and <u>determinate</u> structure. There is usually more than one way of doing this. (Form the <u>primary</u> structure)

(3) Calculate the displacements (deflections) that the known loads cause in the determinate (primary) structure where the redundants have been removed.

(4) Calculate the displacements at these same points in the determinate structure due to the redundants.

(5) At each point where a redundant has been removed, the sum of the displacements calculated in steps (3) & (4) must be equal to the displacement that exists at that point in the actual structure (indeterminate structure). These relationships allow us to evaluate the redundants.

(6) Knowing the values of the redundants, use the equilibrium equations to determine the remaining reactions.

Example (Same as previous example)

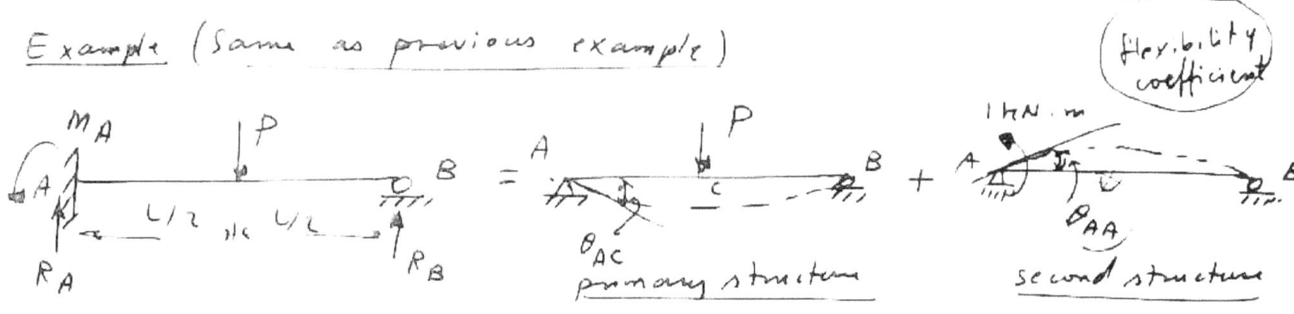

$$\boxed{\theta_A = 0}$$ (no rotation at the fixed end).

$$\boxed{\theta_{AC} + M_A\,\theta_{AA} = 0}$$ equation of consistent deformations.

(1) Primary structure:

Use the moment-area theorems:

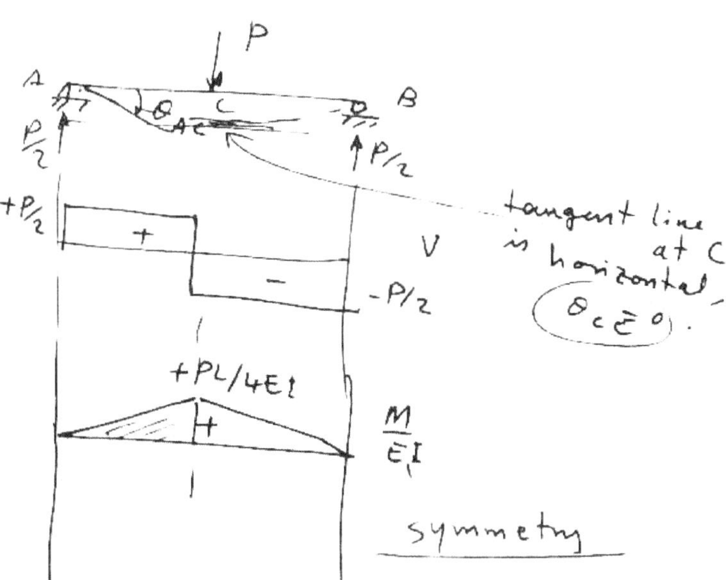

$$\Delta\theta_{AC} = \frac{1}{2}\left(\frac{PL}{4EI}\right)\left(\frac{L}{2}\right)$$

$$\underset{0}{\cancel{\theta_C}} - \theta_{AC} = \frac{PL^2}{16EI}$$

$$\theta_{AC} = \frac{-PL^2}{16EI}$$

(clockwise is negative).

(2) second structure:

$$\theta_{AA} \simeq \tan\theta_{AA} = \frac{\delta_{BA}}{L}$$

$$\delta_{BA} = +\frac{1}{2}(1)(L)\cdot\frac{2}{3}L \cdot \frac{1}{EI} = +\frac{L^2}{3EI}$$

$$\theta_{AA} = +\frac{L^2}{3EI}\cdot\frac{1}{L} = +\frac{L}{3EI}$$

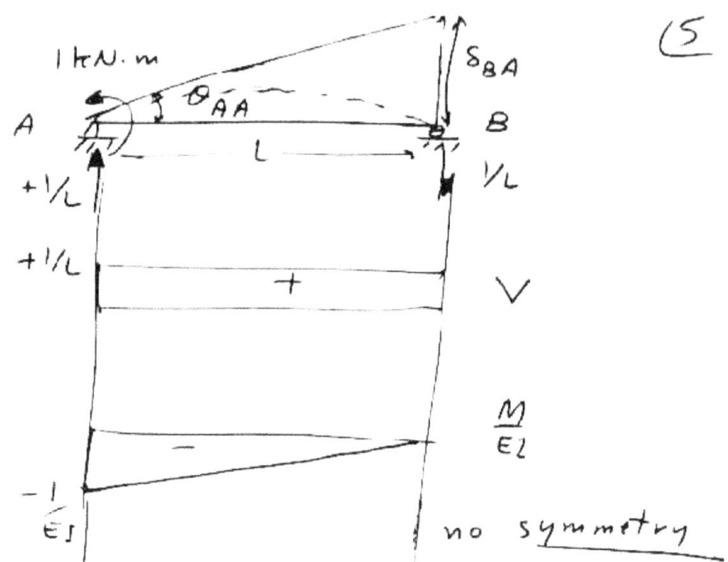

(3) Substitute in the equation of consistent deformations:

$$\theta_{AC} + M_A \theta_{AA} = 0$$

$$-\frac{PL^2}{16EI} + m_A\left(\frac{L}{3EI}\right) = 0 \implies \boxed{M_A = \frac{3}{16}\frac{PL}{}}\ \ \text{counterclockwise}$$

(4) Use the equations of equilibrium to find the other reactions:

$+\circlearrowleft \Sigma M_A = 0$:

$$\frac{3PL}{16} - P\cdot\frac{L}{2} + R_B L = 0$$

$$R_B = \frac{1}{2}P - \frac{3P}{16} = \frac{5P}{16}$$

$+\uparrow \Sigma F_y = 0$: $\quad R_A + \frac{5P}{16} - P = 0 \implies R_A = \frac{11P}{16}$

11.3 Application of Consistent Deformations to Structures with One Redundant:

Example 11.1:

Determine the reactions for the beam shown in the figure and draw its shear and moment diagrams.

Solution:

No. of redundants = 1

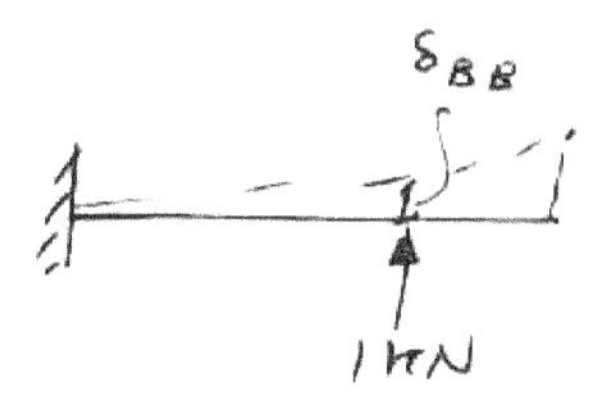

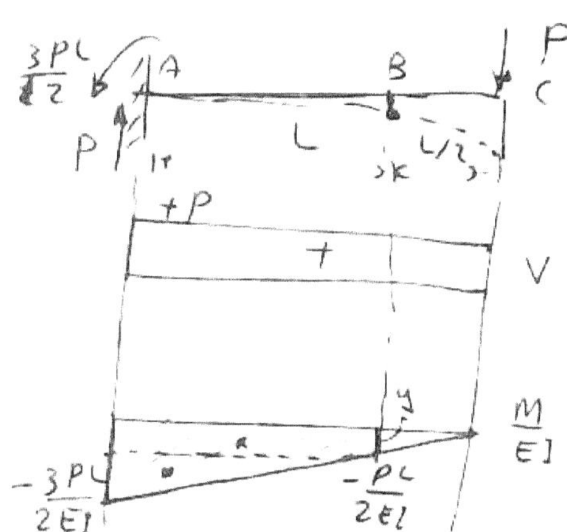

actual structure   primary structure   second structure

$$\delta_B = 0 \implies \boxed{\delta_{BC} + R_B\,\delta_{BB} = 0}\quad \text{equation of consistent deformations.}$$

(1) Primary Structure:

$$y = \frac{L/2}{3L/2} \cdot \frac{3PL}{2EI}$$

$$= \frac{PL}{2EI}$$

$$\delta_{BC} = -\left(\frac{PL}{2EI}\right)(L)\cdot\left(\frac{L}{2}\right)$$

$$\quad -\frac{1}{2}\left(\frac{2PL}{2EI}\right)(L)\cdot\left(\frac{2}{3}\right)(L)$$

$$= -\frac{PL^3}{4EI} - \frac{PL^3}{3EI}$$

$$= -\frac{7PL^3}{12EI}$$

(2) second structure

$$\delta_{BB} = +\frac{1}{2}\left(\frac{L}{EI}\right)(L)\cdot\frac{2}{3}L$$

$$= \frac{L^3}{3EI}$$

(3) Substitute in the equation of consistent deformations:

$$\delta_{BC} + R_B\,\delta_{BB} = 0$$

$$-\frac{7PL^3}{12EI} + R_B\cdot\left(\frac{L^3}{3EI}\right) = 0$$

$$R_B = \frac{7PL^3}{12EI}\cdot\frac{3EI}{L^3} = \frac{7P}{4}\uparrow$$

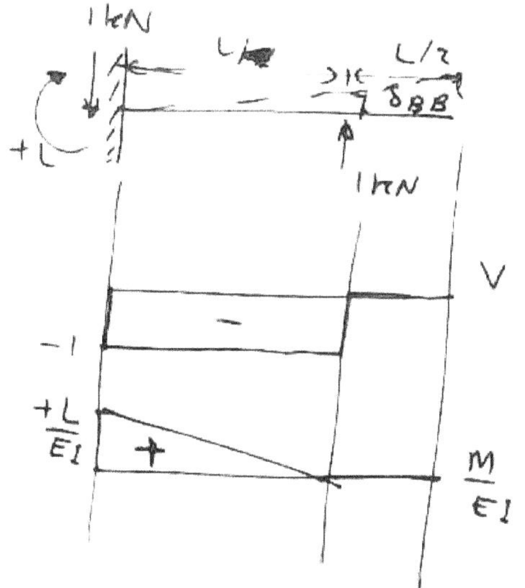

(4) Use the equilibrium equations:

$+\uparrow \Sigma F_y = 0:$

$$R_A + \frac{7P}{4} - P = 0$$

$$R_A = -\frac{3P}{4} \implies R_A = \frac{3P}{4} \downarrow$$

$+\circlearrowleft \Sigma M_A = 0: \quad M_A + \frac{7P}{4}(L) - P\left(\frac{3L}{2}\right) = 0$

$$M_A = \frac{3PL}{2} - \frac{7PL}{4} = -\frac{PL}{4}$$

$$\implies M_A = \frac{PL}{4} \downarrow$$

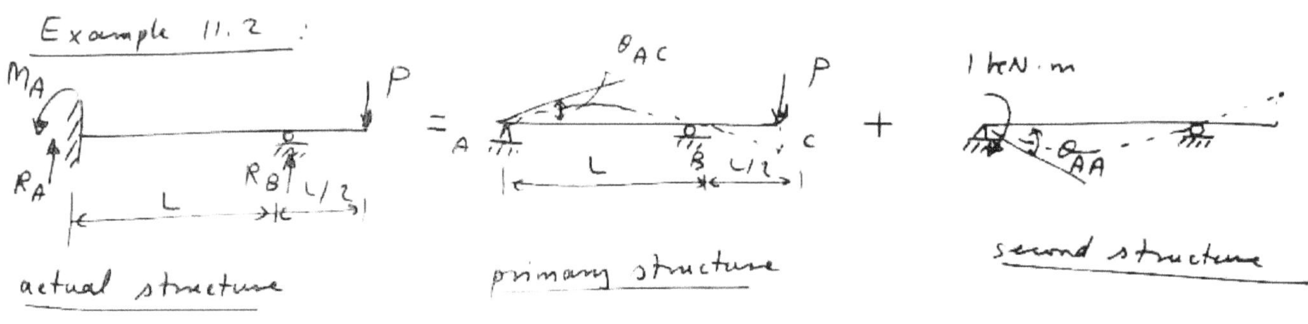

Example 11.2:

actual structure $=$ primary structure $+$ second structure

$$\theta_A = 0 \quad \text{(no rotation at the fixed end)}.$$

$$\boxed{\theta_{AC} + M_A \theta_{AA} = 0} \quad \text{equation of consistent deformations.}$$

(1) primary structure:

$+\circlearrowleft \Sigma M_A = 0: \quad R_B \cdot L - P \cdot \frac{3L}{2} = 0$

$$R_B = \frac{3}{2}P \uparrow$$

$$R_A = \frac{P}{2} \downarrow$$

$$d_1 = \frac{1}{2}\left(\frac{PL}{2EI}\right)(L) \cdot \frac{1}{3}L$$

$$= \frac{PL^3}{12EI}$$

$$\theta_{AC} = \frac{d_1}{L}$$

$$= +\frac{PL^2}{12EI} \quad \text{(counterclockwise)}.$$

(2) second structure

$+\curvearrowright \Sigma M_A = 0$:

$R_B L - 1 = 0$

$R_B = 1/L \uparrow$

$R_A = -1/L \downarrow$

$d_2 = \frac{1}{2}\left(\frac{1}{EI}\right)(L) \cdot \frac{2}{3}L$

$\quad = \dfrac{L^2}{3EI}$

$\delta_{AA} = -\dfrac{d_2}{L} = -\dfrac{L}{3EI}$

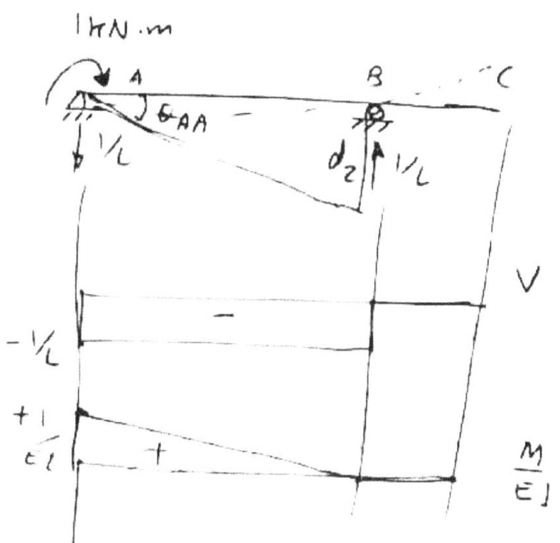

(3) Substitute in the equation of consistent deformations:

$$\delta_{AC} + M_A \delta_{AA} = 0$$

$$\frac{PL^2}{12EI} + M_A\left(-\frac{L}{3EI}\right) = 0 \implies M_A = \frac{PL^2}{12EI} \cdot \frac{3EI}{L} = \frac{PL}{4} \curvearrowright$$

(4) Use the equilibrium equations to find the other reactions:

$+\curvearrowright \Sigma M_A = 0$:

$R_B \cdot L - \dfrac{PL}{4} - P\cdot\left(\dfrac{3L}{2}\right) = 0$

$R_B = \dfrac{1}{4}P + \dfrac{3P}{2} = \dfrac{7P}{4} \uparrow$

$+\uparrow \Sigma F_y = 0:\ R_A + \dfrac{7P}{4} - P = 0 \implies R_A = +\dfrac{3P}{4} \implies R_A = \dfrac{3P}{4} \downarrow$

Example 11.3:

It is required to determine the reactions and to draw the bending-moment diagram for the beam shown in the figure.

Solution:

No. of redundants = 1.

$$\delta_B = 0 \implies \boxed{\delta_{BD} + R_B \delta_{BB} = 0}$$

actual structure = primary structure + second structure

(1) <u>primary structure</u>:

Calculate the reactions
of the conjugate beam:

$+\circlearrowleft \ \Sigma M_A = 0:$

$$R_C(30) - \frac{1}{2}\left(\frac{100}{EI}\right)(20)\left(10 + \frac{20}{3}\right)$$
$$- \frac{1}{2}\left(\frac{100}{EI}\right)(10)\left(\frac{2}{3}\right)(10) = 0$$

$$R_C = \frac{666.7}{EI} \uparrow$$

<u>Section B</u>

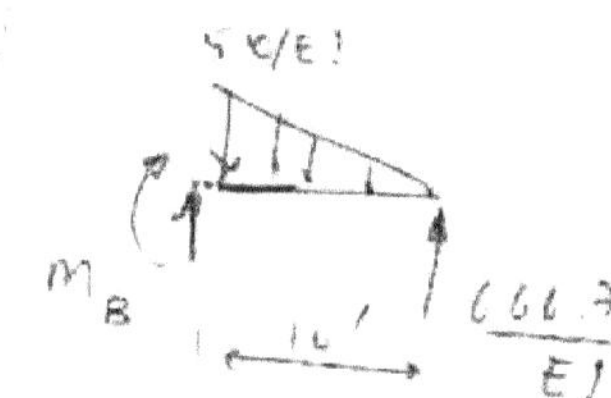

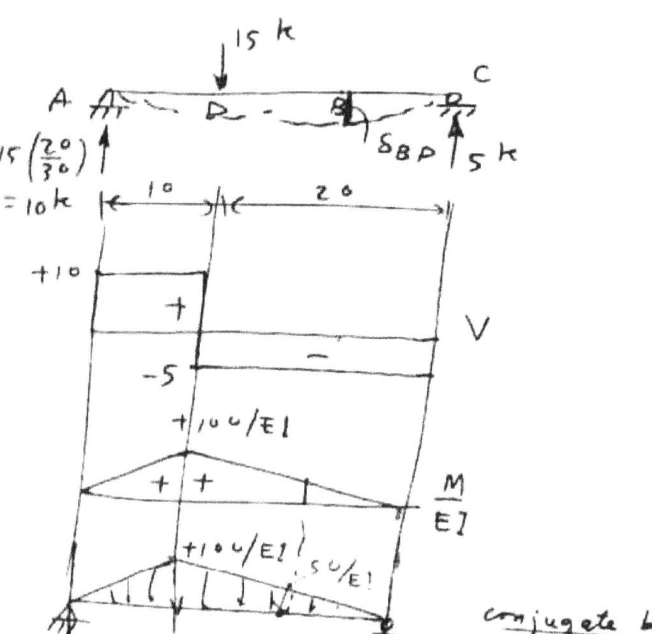

$+\circlearrowleft \ \Sigma M_B = 0:$

$$-m_B - \frac{1}{2}\left(\frac{50}{EI}\right)(10)\left(\frac{10}{3}\right) + \frac{666.7}{EI}(10) = 0$$

$$M_B = \frac{5834}{EI} \implies \delta_{BD} = -\frac{5834}{EI} \ (down).$$

(2) <u>second structure</u>:

Calculate the reactions of
the conjugate beam:

$+\circlearrowleft \ \Sigma M_A = 0:$

$$-R_B(30) + \frac{1}{2}\left(\frac{6.67}{EI}\right)(10)\left(20 + \frac{10}{3}\right)$$
$$+ \frac{1}{2}\left(\frac{6.67}{EI}\right)(20)\left(\frac{2}{3}\right)(20) = 0$$

$$R_B = \frac{55.6}{EI} \downarrow.$$

<u>Section B</u>:

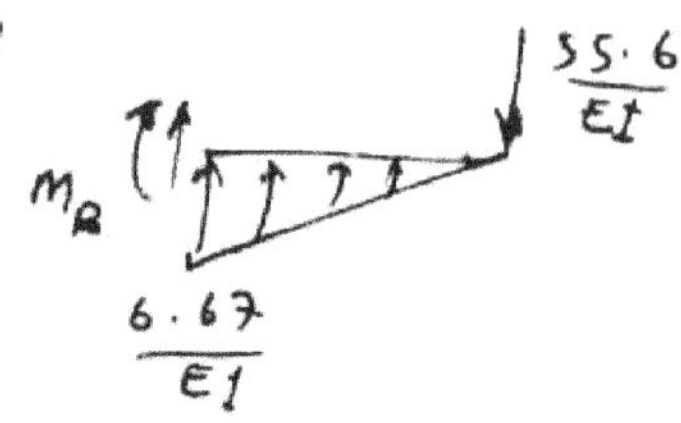

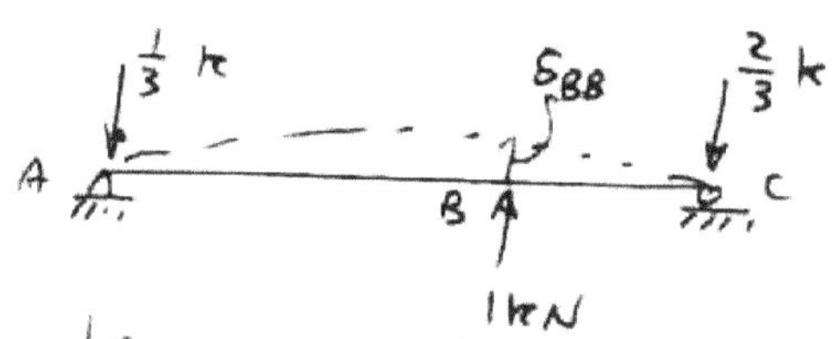

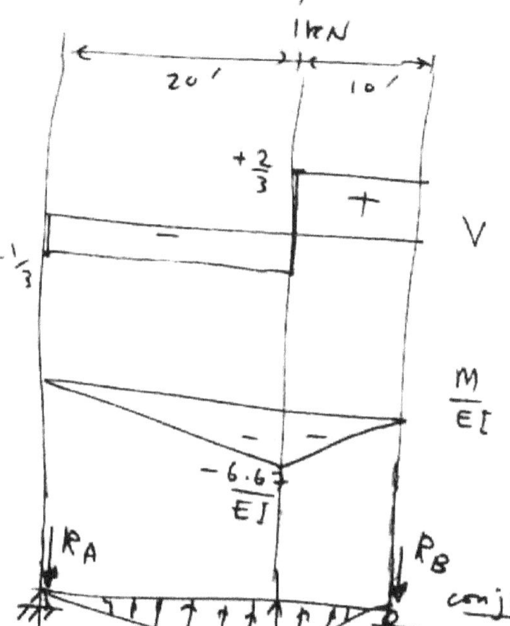

$$M_B = \frac{1}{2}\left(\frac{6.67}{EI}\right)(10)\left(\frac{1}{3}\right)(10) - \frac{55.6}{EI}(10)$$
$$= -\frac{444.8}{EI} \implies \delta_{BB} = \frac{444.8}{EI}.$$

(3) Substitute in the equation of consistent deformations:

$$\delta_{BD} + R_B \,\delta_{BB} = 0$$

$$-\frac{5834}{EI} + R_B\left(\frac{444.8}{EI}\right) = 0$$

$$R_B = 13.1 \ k \uparrow .$$

(4) Use the equilibrium equations to find the other reactions.

$$+\circlearrowright \ \Sigma M_A = 0:$$

$$-15(10) + 13.1(20) + R_c(30) = 0$$

$$R_c = -3.73 \ k \implies R_c = 3.73^k \downarrow .$$

$$+\uparrow \ \Sigma F_y = 0:$$

$$R_A - 15 + 13.1 + 3.73 = 0$$

$$R_A = 5.63k\uparrow \implies$$

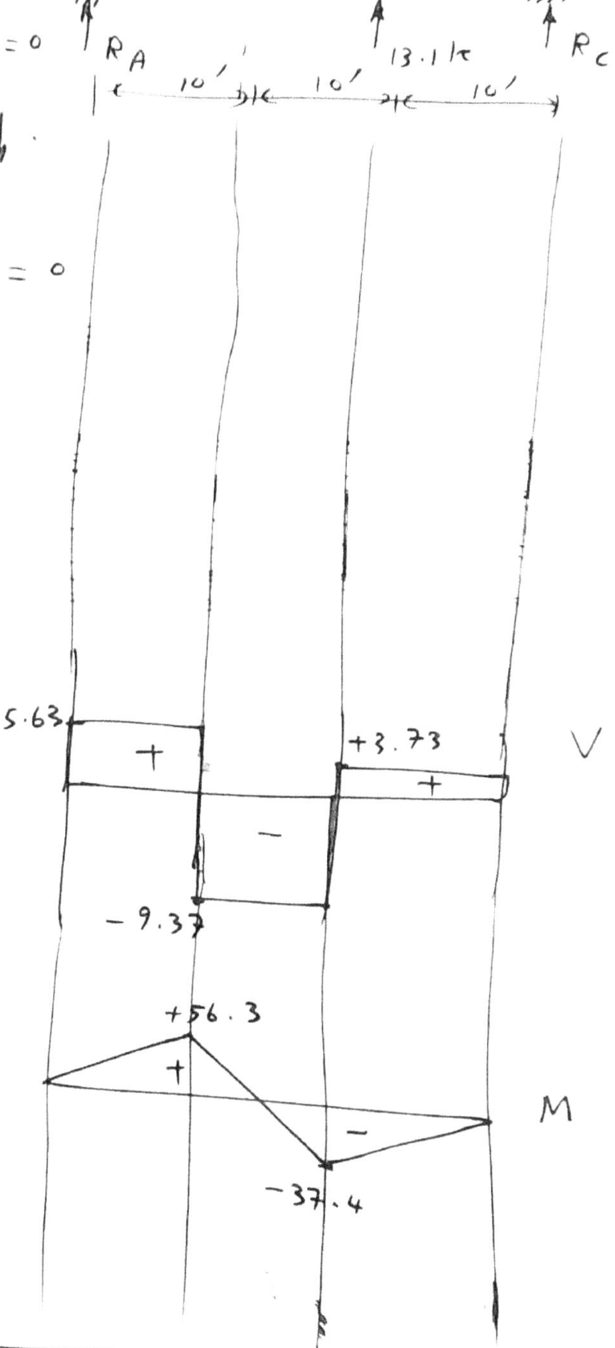

Example 11.4 :

It is required to determine the reactions
and to draw the bending-moment
diagram for the frame shown in the
figure. Both members of the frame
are assumed to have the same EI.

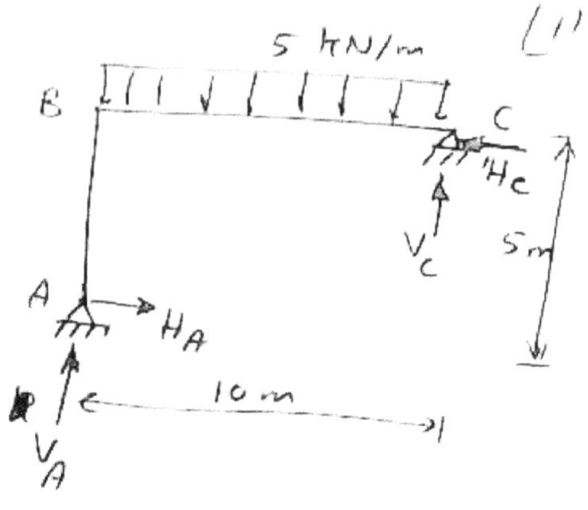

Solution : No. of redundants = 1

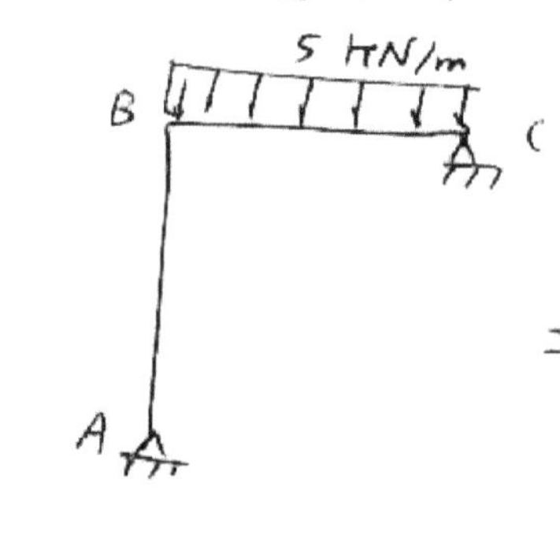

actual structure

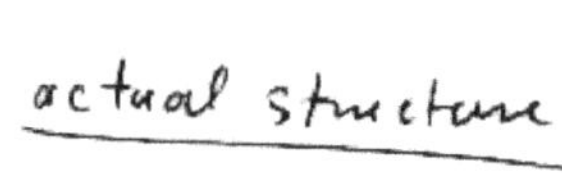

=

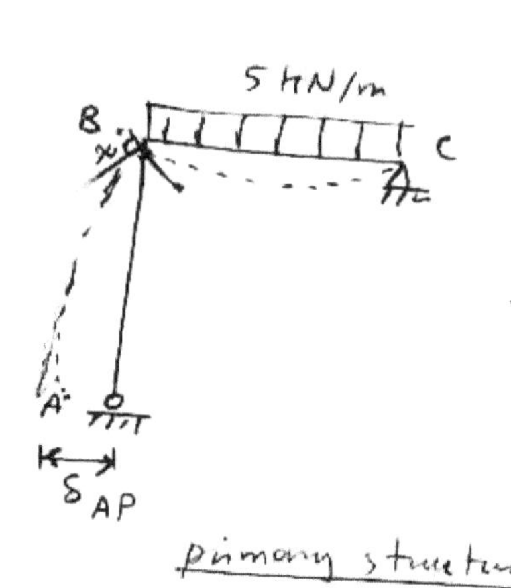

primary structure

+

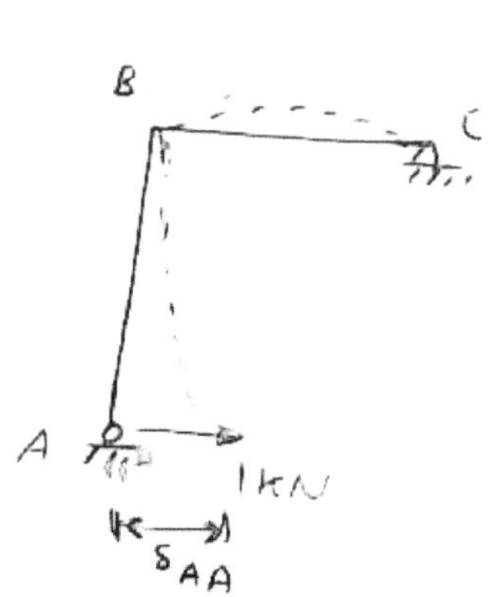

second structure
(dummy structure)

$\delta_{A_h} = 0$ $\Rightarrow$ $\boxed{\delta_{AP} + H_A \delta_{AA} = 0}$ equation of consistent deformate

(1) primary structure :
use the second structure
as the dummy frame.

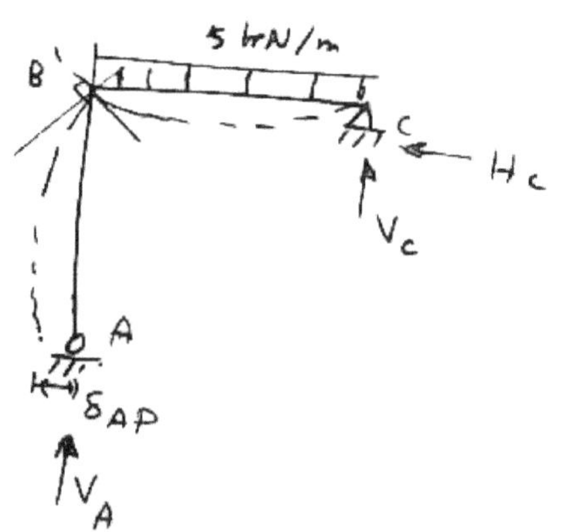

$+\circlearrowright \Sigma M_c = 0$:

$5(10)(5) - V_A(10) = 0$

$V_A = 25$ k $\uparrow$.

$\stackrel{+}{\rightarrow} \Sigma F_x = 0$: $H_c = 0$.

$+\uparrow \Sigma F_y = 0$: $V_A + 25 - 5(10) = 0$

$V_A = 25$ k $\uparrow$.

dummy structure :

$V_A = V_c = \dfrac{(1)(5)}{10} = 0.5$ k .

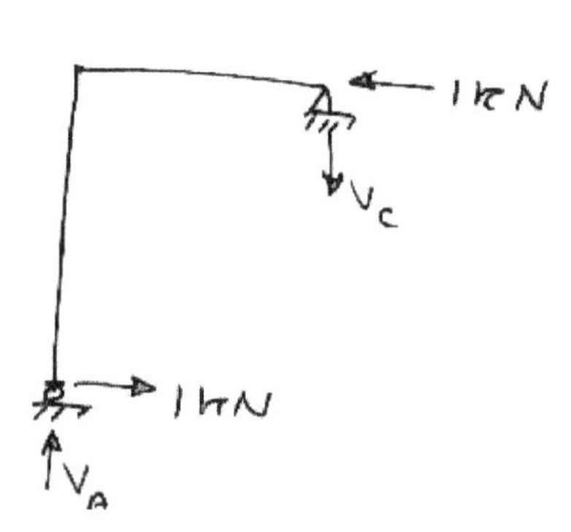

Member AB

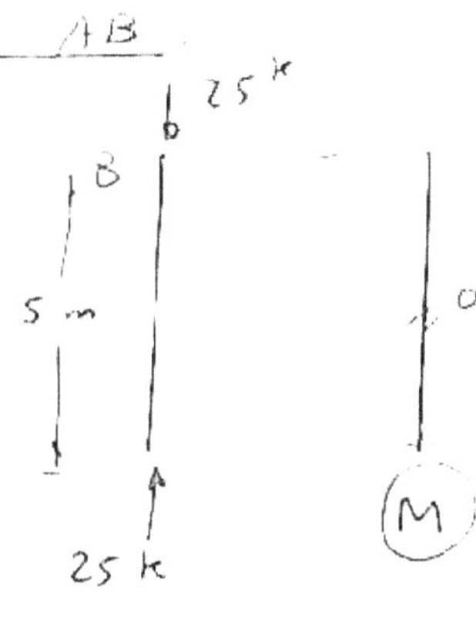

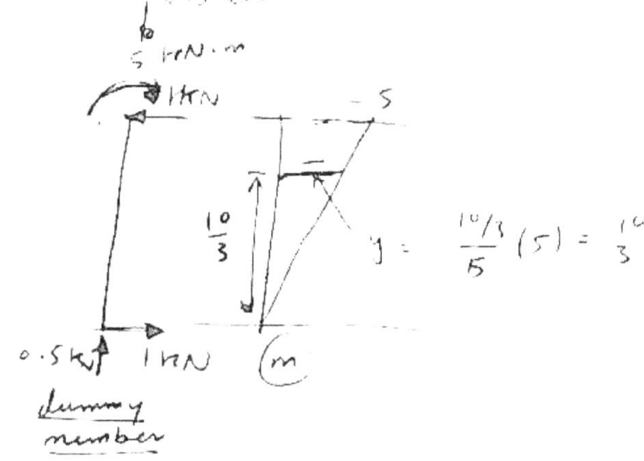

dummy member

Member BC:

5 kN/m

10 m

25 k   25 k

$\dfrac{5(10)^2}{8} = 62.5$ kN·m

(M)

0.5 kN

5 kN·m

1 kN →   ↑ dummy member
0.5 kN   1 kN

0.5   +0.5   v

$\dfrac{20}{3}$

(m)

−5   −2.5 $\dfrac{20}{3}(5) = \dfrac{100}{3}$

$\dfrac{1}{EJ}\displaystyle\int_0^L \dfrac{Mm}{EI}\,dx$

$\therefore\ \delta_{AP} = \delta_{AP(1)} + \delta_{AP(2)} \qquad = 0 \qquad -\dfrac{1}{EI}\left[2\left(\dfrac{2}{3}\right)(62.5)(5)\right](2.5)$

$$= -\dfrac{1041.7}{EI}$$

$$\delta_{AP} = -\dfrac{1041.7}{EI} \quad (\text{to the left}).$$

(2) second structure. (the same as the dummy structure).

$$\delta_{AA} = \delta_{AA(1)} + \delta_{AA(2)} = \dfrac{1}{EJ}\int_0^L \dfrac{mm}{EI}\,dx$$

$$= \dfrac{1}{EI}\left[-\dfrac{1}{2}(5)(5)\right]\left(-\dfrac{10}{3}\right) + \dfrac{1}{EI}\left[-\dfrac{1}{2}(5)(10)\right]\left(-\dfrac{10}{3}\right)$$

$$= +\dfrac{125}{EI}.$$

(3) substitute in the equation of consistent deformations : $\qquad$ 13

$$\delta_{AP} + H_A \, \delta_{AA} = 0$$

$$-\frac{1041.7}{EI} + H_A \left(\frac{125}{EI}\right) = 0 \implies H_A = \underline{8.33 \text{ kN} \longrightarrow}$$

(4) Use the equilibrium equations to determine the other reactions.

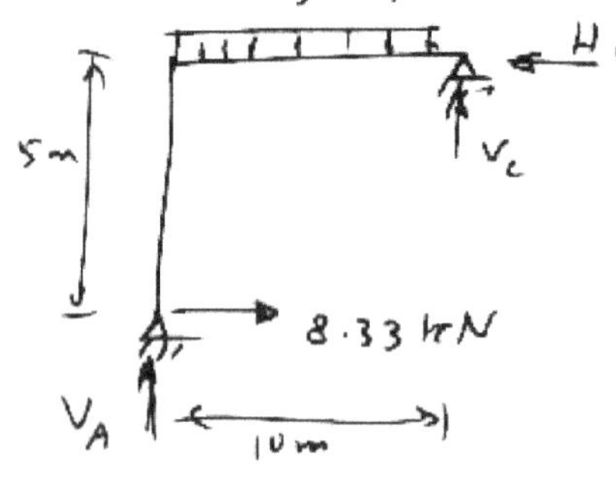

$\xrightarrow{+} \Sigma F_x = 0$: $\quad H_c = 8.33 \text{ kN} \longleftarrow$

$+\circlearrowleft \Sigma M_A = 0$: $\quad V_c(10) - 5(10)(5)$
$\qquad\qquad + 8.33(5) = 0$

$\qquad V_c = 20.84 \text{ kN} \uparrow$

$+\uparrow \Sigma F_y = 0$: $V_A + 20.84 - 50 = 0$

$\qquad V_A = 29.16 \text{ kN} \uparrow$

<u>B.M.D. :</u>

<u>Member AB :</u>

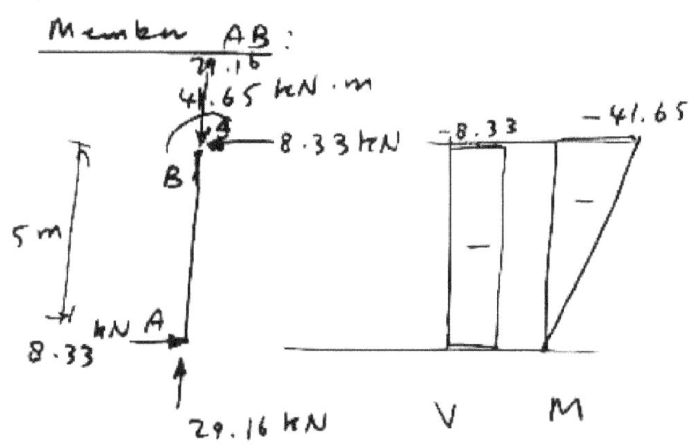

<u>Member BC :</u>

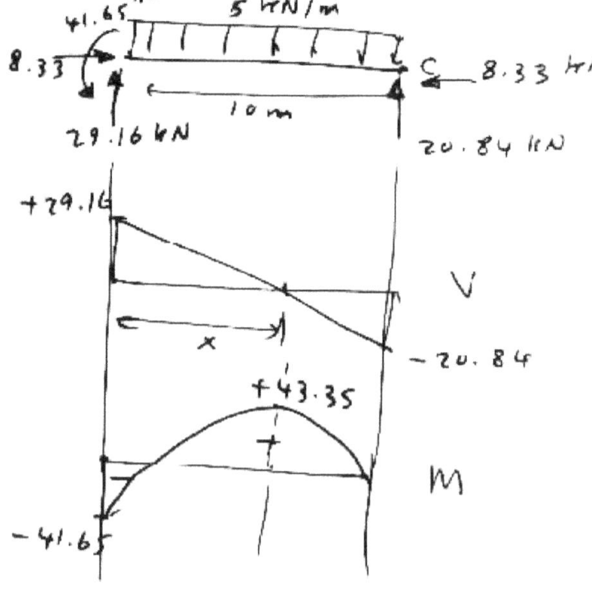

Check :

$$M_c = 41.65 + 5(10)(5) - 29.16(10)$$
$$= 0 \quad \checkmark$$

Locate the point of zero shear :

$$\frac{x}{29.16} = \frac{10-x}{20.84}$$

$$291.6 - 29.16 x = 20.84 x$$

$$\implies x = \underline{\underline{5.83 \text{ m}}} .$$

$$M_{max} = -41.65 + \tfrac{1}{2}(29.16)(5.83) = 43.35 \text{ kN·m}$$

Example : Repeat the previous example with the following data.

for member BC : EI .

for member AB : $0.5\, EI = \dfrac{EI}{2}$.

Solution :

$$\delta_{AP} = \frac{-1041.7}{EI}$$

$$\delta_{AA} = \frac{2}{EI}\left[\frac{1}{2}(5)(5)\left(\frac{10}{3}\right)\right] + \frac{1}{EI}\left[\frac{1}{2}(5)(10)\left(\frac{10}{3}\right)\right]$$

$$= \frac{166.7}{EI}$$

Substitute in $\delta_{AP} + H_A\,\delta_{AA} = 0$

$$\frac{-1041.7}{EI} + H_A\left(\frac{166.7}{EI}\right) = 0$$

$$H_A = 6.25 \ kN \longrightarrow$$

Calculate the other reactions using the equations of equilibrium.

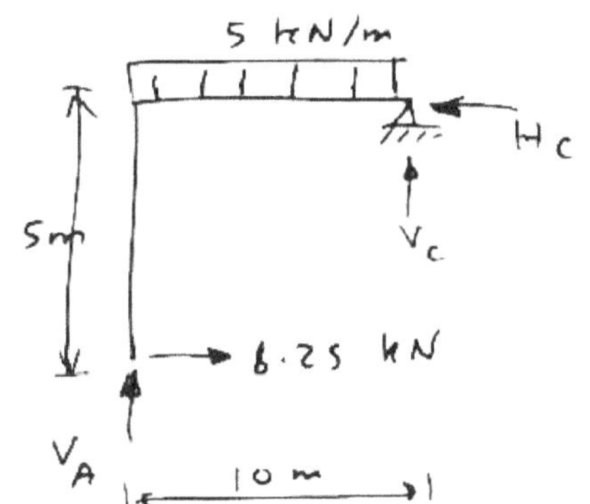

$\xrightarrow{\;+\;}\ \Sigma F_x = 0 : \ H_c = 6.25 \ kN \ \longleftarrow$

$+\circlearrowright\ \Sigma M_A = 0 : \ V_c(10) + 6.25(5) - 5(10)(5) = 0$

$$V_c = 21.88 \ kN \uparrow$$

$+\uparrow\ \Sigma F_y = 0 : \ V_A + 21.88 - 50 = 0$

$$V_A = 28.12 \ kN \uparrow$$

B.M.D. :

Member AB :

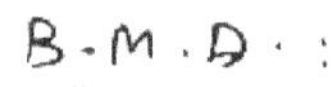

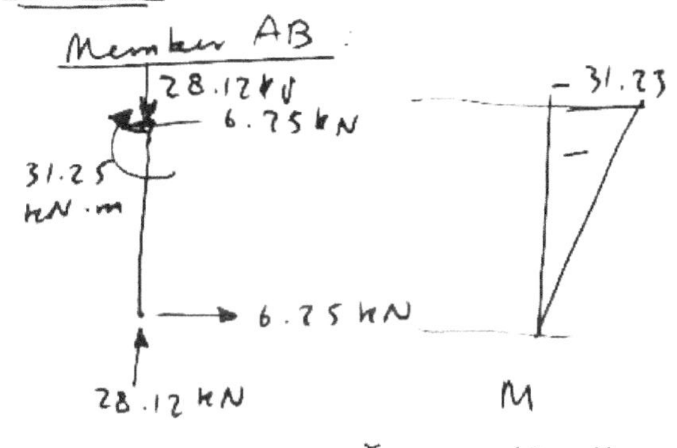

Member BC :

Determine x : $\dfrac{x}{28.12} = \dfrac{10-x}{21.88}$

$$21.88x = 281.2 - 28.12x \implies x = 5.62\,m$$

$$M_{max} = -31.25 + \frac{1}{2}(28.12)(5.62) = 46.82\ kN.m$$

Example 11.5 :

Beam : $I = 100 \times 10^6 \ mm^4$

$E = 200 \times 10^6 \ kN/m^2$

Cable : $A = 20 \ mm^2$

$E = 200 \times 10^6 \ kN/m^2$.

Calculate the reactions.

Solution : No. of redundants = 1

Consider the tension T in the cable as the redundant.

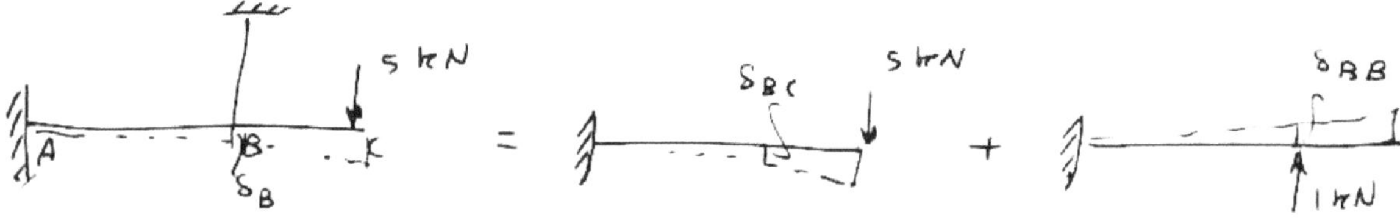

actual structure   primary structure   second structure

$$\delta_B = \boxed{\delta_{BC} + T\,\delta_{BB} = \frac{TL}{EA}} \quad \text{(elongation in the cable)}$$
equation of consistent deformation

(1) primary structure :

$$\delta_{BC} = \frac{15}{EI}(6)(3) + \frac{1}{2}\left(\frac{30}{EI}\right)(6)(4)$$

$$= \frac{630}{EI} \quad (down)$$

take $\delta_{BC} = -\frac{630}{EI}$.

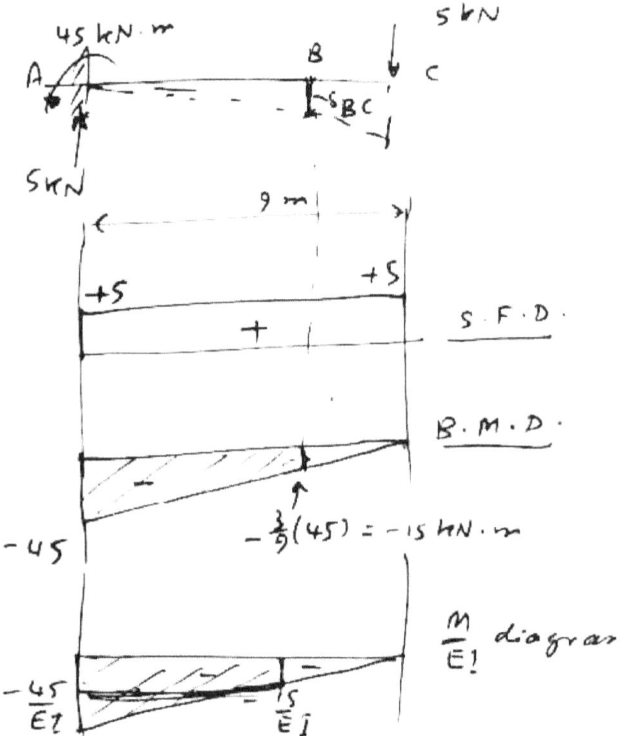

(=) second structure:

$$\delta_{BB} = \frac{1}{2}\left(\frac{6}{EI}\right)(6)(4) = \frac{72}{EI} \quad (up)$$

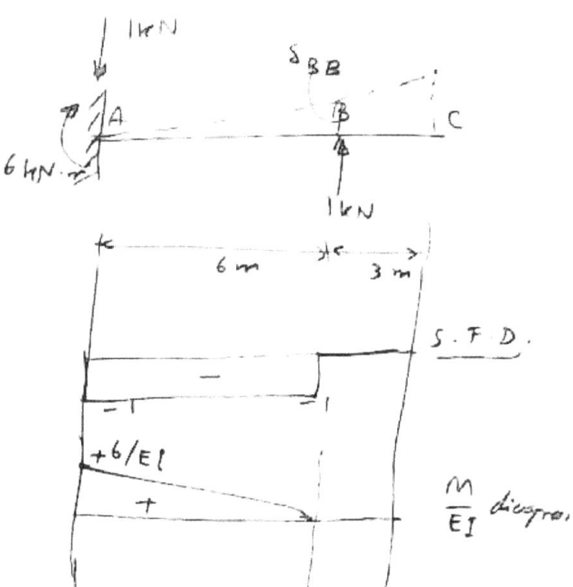

(3) Equation of Consistent Deformation:
(Compatibility Equation)

$$\delta_{BC} + T\delta_{BB} = -\frac{TL}{EA} \quad (down) \quad (table)$$

$$-\frac{630}{EI} + T\left(\frac{72}{EI}\right) = -\frac{T(10)}{EA}$$

$$\left(\frac{72T}{I} + \frac{10T}{A}\right) = \frac{630}{I}$$

$$T\left(\frac{72}{I} + \frac{10}{A}\right) = \frac{630}{I}$$

$$T\left[\frac{72}{100\times10^6 \times(10^{-3})^4} + \frac{10}{20\times(10^{-3})^2}\right] = \frac{630}{100\times10^6 (10^{-3})^4}$$

$$T\left(\frac{72}{100}\times10^6 + \frac{10}{20}\times10^6\right) = \frac{630}{100}\times10^6$$

$$T\left(\frac{72}{100} + \frac{1}{2}\right) = 6.3 \implies T(0.72 - 0.5) = 6.3$$

$$T = \frac{6.3}{0.72+0.5} = 5.16 \text{ kN}$$

(4) Use the equations of equilibrium to find the other reactions:

$$+\uparrow \; \Sigma F_y = 0: \; -A_y + 5.16 - 5 = 0$$

$$A_y = 0.16 \text{ kN} \downarrow$$

$$+\circlearrowright \; \Sigma M_A = 0: \; -M_A - 5(9) + 5.16(6) = 0$$

$$M_A = -14.04 \text{ kN·m}$$

$$\therefore M_A = 14.04 \text{ kN·m} \circlearrowright$$

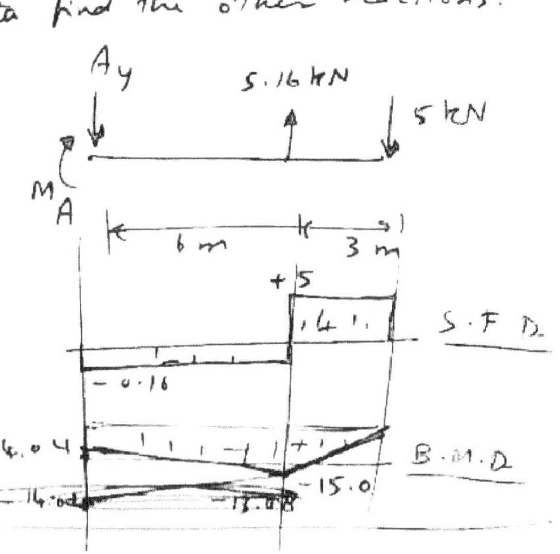

Example 11.5 :

Calculate the reactions at the fixed end and the tension in the cable for the <u>composite structure</u> shown in the figure.

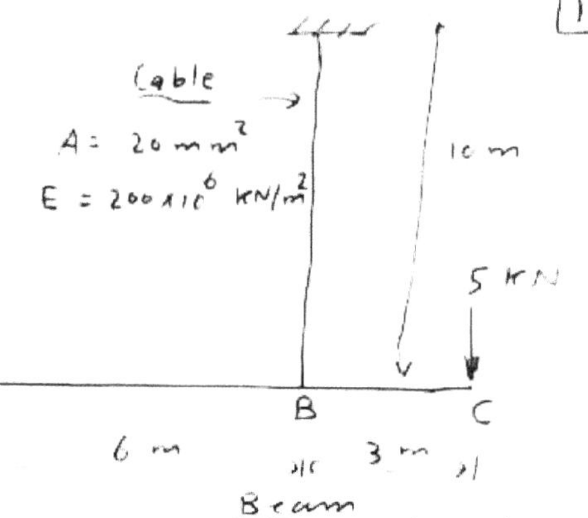

<u>Solution</u> :

No. of Unknowns = 4
No. of Equations = 3

∴ The structure is indeterminate to the first degree (one redundant).

Use the method of consistent deformations.

Choose T as the redundant.

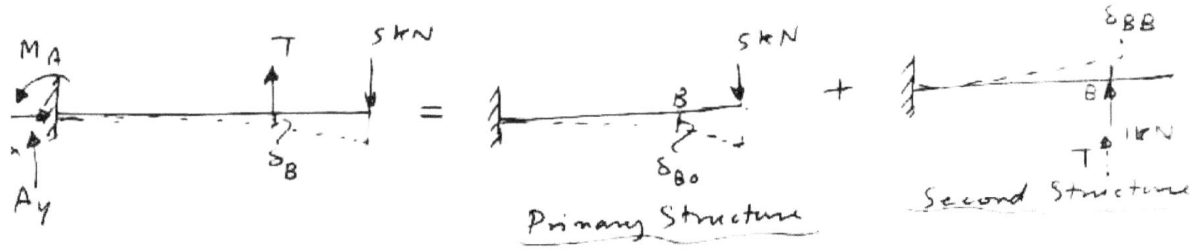

<u>Note</u> : δ_{BB} is called the <u>Flexibility coefficient</u>, and is denoted by $\underline{f_{BB}}$.

Equation of Consistent Deformation (Compatibility Equation):

$$\boxed{\delta_B = \delta_{B0} + T\,\delta_{BB}} \quad \underset{=}{or} \quad \boxed{\delta_B = \delta_{B0} + T f_{BB}}$$

<u>Step 1</u>. Calculate δ_{B0} from the primary structure.
Use the moment-area theorems.

δ_{B0} = moment of $\frac{M}{EI}$ diagram between A and B about B

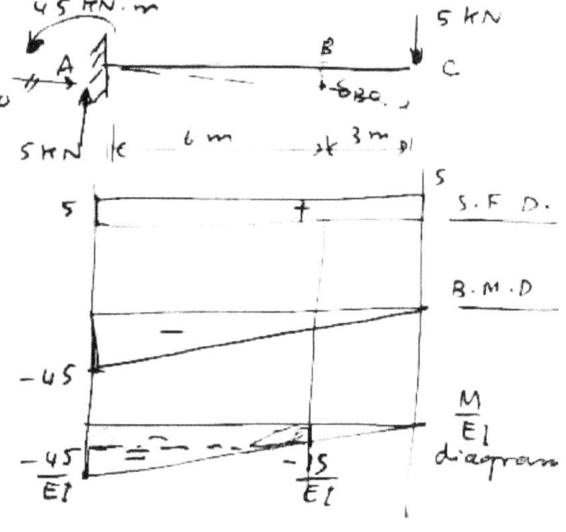

$$= -\frac{15}{EI}(6)(3) - \frac{1}{2}\left(\frac{30}{EI}\right)(6)(4)$$

$$\delta_{B0} = -\frac{630}{EI} \qquad (down)$$

Step 2   Calculate f_{BB} from the second structure.

f_{BB} = moment of $\frac{M}{EI}$ diagram
between A and B about B

$$= \frac{1}{2}\left(\frac{6}{EI}\right)(6)(4)$$

$$f_{BB} = +\frac{72}{EI} \quad (up)$$

Step 3 : Equation of Consistent Deformation

$$\delta_B = \delta_{BO} + T f_{BB}$$

In this case, $\delta_B \neq 0$

$$\delta_B = \frac{-TL}{EA} \quad (\text{extension in the cable}) \cdot (\underline{down})$$

$$-\frac{TL}{E_c A} = -\frac{630}{E_b I} + T\left(\frac{72}{E_b I}\right) \quad , \quad \text{but} \quad E_b = E_c = E$$
$$\hspace{7cm} (E \text{ cancels})$$

$$\therefore \quad -\frac{T(10)}{20\times(10^{-3})^2} = \frac{-630}{100\times10^{6}\times(10^{-3})4} + T\left(\frac{72}{100\times10^{6}\times(10^{-3})4}\right)$$

$$\therefore \quad 1,220,000\ T = 6,300,000 \quad \Rightarrow \quad \underline{T = 5.16\ kN}$$

Step 4 : Equations of Equilibrium :

$\xrightarrow{+} \ \Sigma F_x = 0 : \quad A_x = 0 .$

$+\uparrow \ \Sigma F_y = 0 : \quad A_y + 5.16 - 5 = 0$

$$A_y = -0.16\ kN$$
$$A_y = 0.16\ kN \ \downarrow$$

$+\circlearrowleft \ \Sigma M_A = 0 : \quad M_A + 5.16(6) - 5(9) = 0$

$$M_A = 14.04\ kN \cdot m \ \circlearrowleft$$

<u>Note</u> : If the support at B had been <u>rigid</u>, i.e. $\delta_B = 0$, then

$$\delta_B = 0 = \delta_{BO} + R_B f_{BB}$$

$$0 = -\frac{630}{E_b I} + T\left(\frac{72}{E_b I}\right) \quad \Rightarrow \quad \underline{R_B = 8.75\ kN \ \uparrow}$$

The rigid support takes more load than the elastic support

11·4 Support Settlement.

* In general, the problem of support settlement is far more serious for indeterminate structures than it is for determinate ones.

Example 11·6 :

1) Determine the reactions and draw the bending moment diagram for the <u>continuous beam</u> shown in the figure.

2) Compare the results obtained with those corresponding to zero settlement.

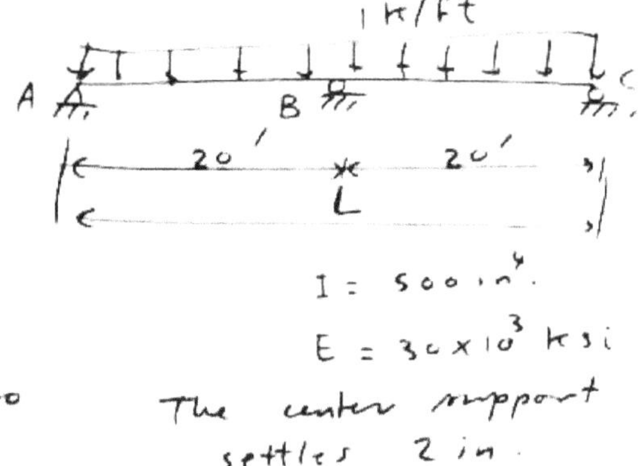

<u>Solution</u> :

① Use the method of consistent deformations.
The beam is indeterminate to the first degree (one redundant)

Choose R_B as the redundant.

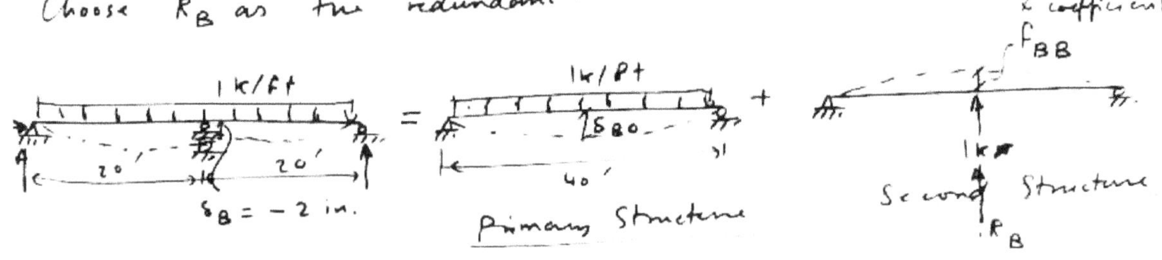

$$\delta_B = \delta_{BO} + R_B f_{BB}$$

<u>Step 1</u> : Calculate δ_{BO} from the primary structure.
Use the moment area theorems.

$$\delta_{BO} = -\frac{2}{3}\left(\frac{200}{EI}\right)(20)\left(\frac{5}{8}\right)(20)$$

$$= -\frac{100,000}{3EI} \quad (down)$$

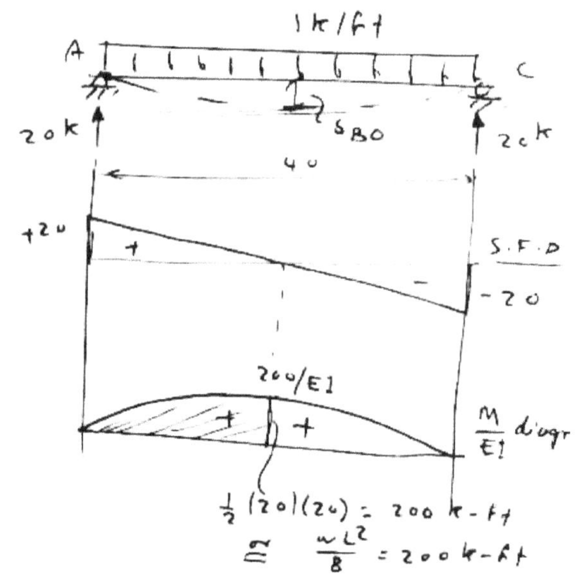

Step 2 : Calculate f_{BB} from the second structure

$$f_{BB} = +\left(\frac{10}{EI}\right)(20)\left(\frac{1}{2}\right)\left(\frac{2}{3}\right)(20)$$

$$= + \frac{4000}{3EI} \qquad (up)$$

Step 3 : Equation of Consistent Deformation.

$$\delta_B = \delta_{BO} + R_B f_{BB}$$

$$-\frac{2}{12} = \frac{-100,000}{3(30\times10^3)(500)}(12)^2 + R_B\left(\frac{4000}{3(30\times10^3)(500)}(12)^3\right)$$

$$-\frac{1}{6} = \frac{(-100,000 + 4,000 R_B)}{3(30\times10^3)(500)}(12)^2$$

$$-5,208.3333 = -100,000 + 4,000 R_B$$

$$\underline{\underline{R_B = 11.98 \ k \uparrow}}$$

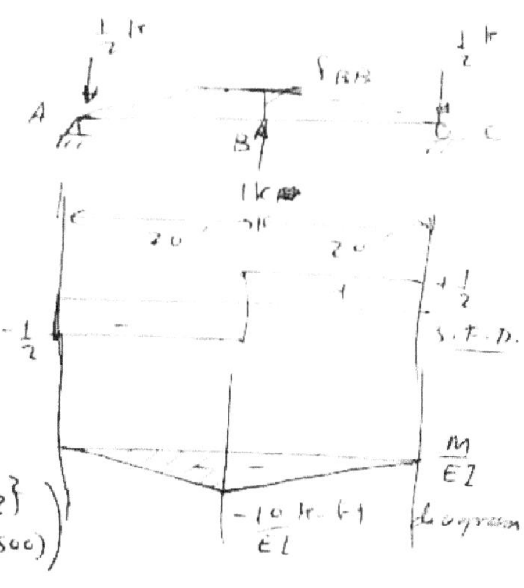

Step 4 : Equations of Equilibrium :

$\xrightarrow{+} \Sigma F_x = 0 : \quad A_x = 0$

$\Sigma M_A = 0 :$

$\quad C_y(40) + 11.98(20) - 1(40)(20) = 0$

$\quad C_y = 14.01 \ k \uparrow$

$+\uparrow \Sigma F_y = 0 :$

$\quad A_y + 11.98 + 14.01 - 1(40) = 0$

$\quad A_y = 14.01 \ k \uparrow$

Locate the point of zero shear :

$$\frac{x}{14.01} = \frac{20-x}{5.99}$$

$$5.99 x = 20(14.01) - 14.01 x$$

$$\Rightarrow x = 14.01 \ ft.$$

$$\therefore M_o = \frac{1}{2}(14.01)(14.01) = 98.14 \ k\text{-}ft.$$

$$M_B = 98.14 - \frac{1}{2}(5.99)(20-14.01) = 80.20 \ k\text{-}ft$$

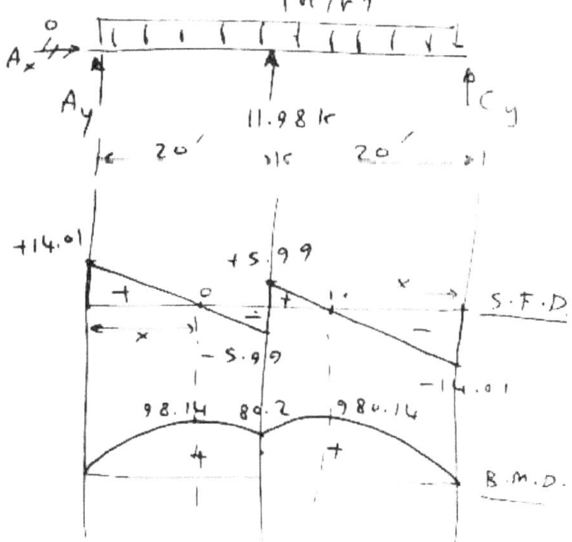

(2) In case there is zero settlement, $\delta_B = 0$:

$$\delta_B = 0 = \delta_{B0} + R_B f_{BB}$$

$$0 = -\frac{100,000}{3EI} + R_B\left(\frac{4000}{3EI}\right) \quad \text{(larger reaction)}$$

$$\therefore R_B = 25\ k \uparrow$$

Equilibrium equations:

$+\circlearrowleft \ \Sigma M_A = 0$:

$$C_y(40) + 25(20) - 1(40)(20) = 0$$

$$C_y = 7.5\ k \uparrow$$

$+\uparrow \ \Sigma F_y = 0$:

$$A_y + 25 + 7.5 - 1(40) = 0$$

$$A_y = 7.5\ k \uparrow$$

Locate the point of zero shear:

$$\frac{x}{7.5} = \frac{20-x}{12.5}$$

$$17.5\,x = 20(7.5) - 7.5\,x$$

$$x = 7.5\ ft.$$

$$M_0 = \tfrac{1}{2}(7.5)(7.5) = 28.125\ k\text{-}ft.$$

$$M_B = 28.125 - \tfrac{1}{2}(12.5)(20-7.5) = -50\ k\text{-}ft.$$

---

* Comparing the bending moment diagrams for the two cases, it is clear that support settlement can give rise to significant increases in stress in an indeterminate structure.

In this example, $M_{max.}$ increased from 50 k-ft to 98.14 k-ft.

11.5 Structures with Several Redundants:

* Use numerical subscripts instead of letters.
* The deflections are measured in the directions of the redundant.

Example :

Use the method of
consistent deformation.

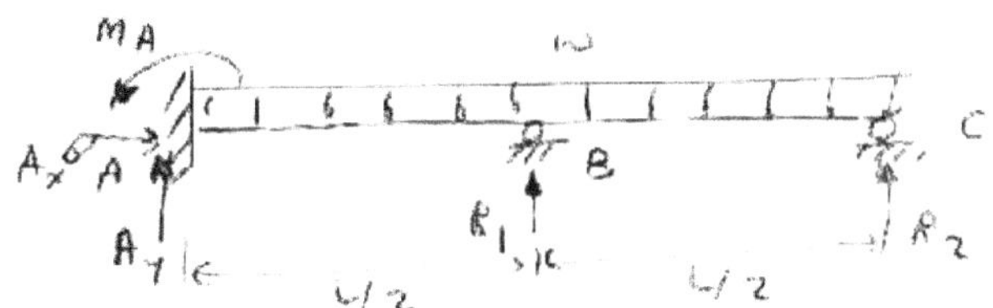

The beam is indeterminate to
the second degree (two redundants).

Choose R_B and R_C as the redundants.
Use R_1 and R_2 (numerical subscripts).

Equations of Consistent Deformations:

$$\delta_1 = 0 = \delta_{10} + R_1 f_{11} + R_2 f_{12} \quad —(1)$$

$$\delta_2 = 0 = \delta_{20} + R_1 \delta_{21} + R_2 \delta_{22} \quad —(2)$$

We get two simultaneous equations:

Rewrite the two equations in

matrix form :

$$\begin{bmatrix} f_{11} & f_{12} \\ f_{21} & f_{22} \end{bmatrix} \begin{Bmatrix} R_1 \\ R_2 \end{Bmatrix} + \begin{Bmatrix} \delta_{10} \\ \delta_{2c} \end{Bmatrix} = 0$$

$$\begin{bmatrix} f_{11} & f_{12} \\ f_{12} & f_{22} \end{bmatrix} \begin{Bmatrix} R_1 \\ R_2 \end{Bmatrix} = - \begin{Bmatrix} \delta_{10} \\ \delta_{20} \end{Bmatrix}$$

The matrix $\begin{bmatrix} f_{11} & f_{12} \\ f_{12} & f_{22} \end{bmatrix}$ is called the <u>flexibility</u> matrix.

Here, we used $f_{12} = f_{21}$ by Maxwell's Law of Reciprocal Deflection

---

* f_{ij} is called the <u>flexibility influence coefficient</u>

  or just the <u>flexibility coefficient</u>

Example 11.7 :

Draw the bending moment diagram for the <u>continuous beam</u> shown in the figure.

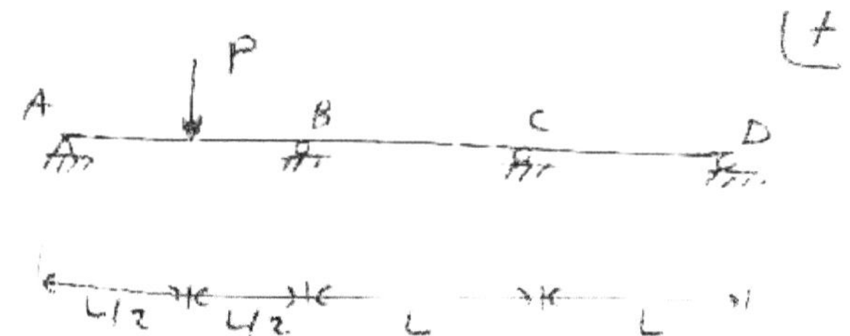

Solution :

The beam is indeterminate to the second degree (two redundants).

Note that $f_{12} = f_{21}$

Equations of Consistent Deformation:

$$\delta_1 = 0 = \delta_{10} + R_1 f_{11} + R_2 f_{12}$$
$$\delta_2 = 0 = \delta_{20} + R_1 f_{21} + R_2 f_{22}$$

$$\begin{bmatrix} f_{11} & f_{12} \\ f_{12} & f_{22} \end{bmatrix} \begin{Bmatrix} R_1 \\ R_2 \end{Bmatrix} = - \begin{Bmatrix} \delta_{10} \\ \delta_{20} \end{Bmatrix}$$

↗ <u>Flexibility matrix</u>

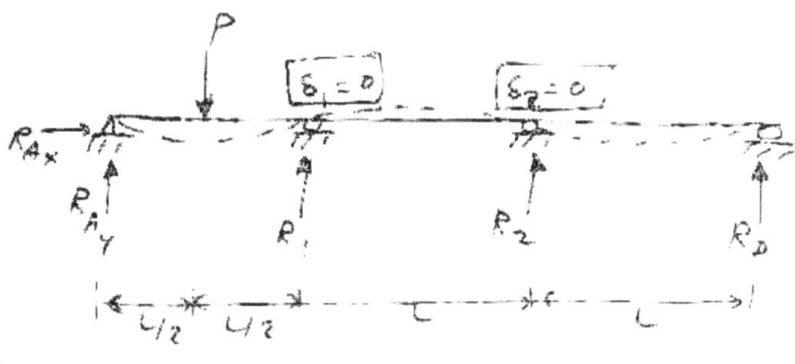

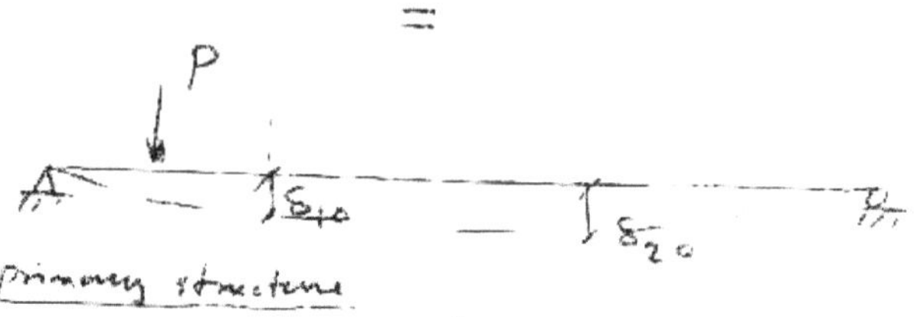

primary structure

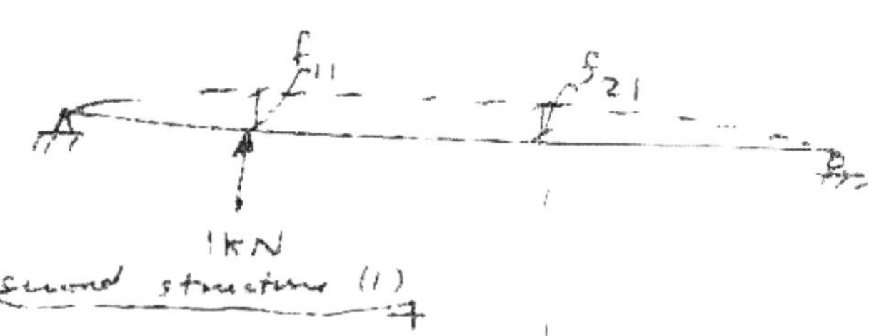

1 kN

second structure (1)

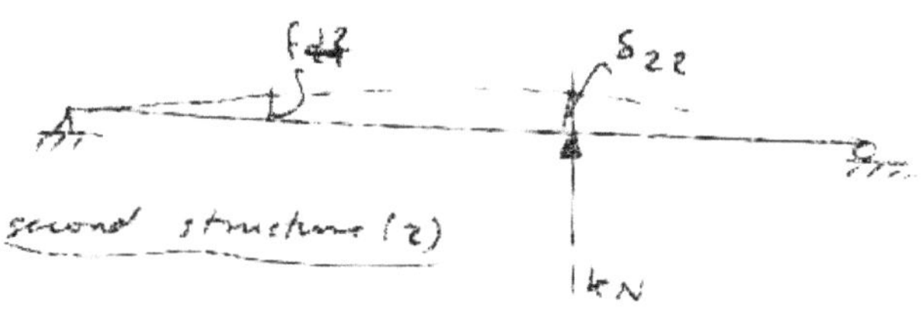

second structure (2)

1 kN

Step 1 : Calculate δ_{10} and δ_{20} from the primary structure.

Use the conjugate beam method.

$$R_A = \frac{\frac{5L}{2}}{3L} P = \frac{5}{6} P .$$
$$R_D = \frac{\frac{L}{2}}{3L} P = \frac{P}{6} .$$
$$M_P = \frac{5}{6} P \cdot \frac{L}{2} = \frac{5}{12} PL .$$

For the conjugate beam.

$+ \circlearrowleft \sum M_A = 0 :$

$$R_D'(3L) - \frac{1}{2}\left(\frac{5PL}{12EI}\right)\left(\frac{5L}{2}\right)\left(\frac{L}{2} + \frac{1}{3}\left(\frac{5L}{2}\right)\right)$$
$$- \frac{1}{2}\left(\frac{5PL}{12EI}\right)\left(\frac{L}{2}\right)\left(\frac{2}{3}\right)\left(\frac{L}{2}\right) = 0$$

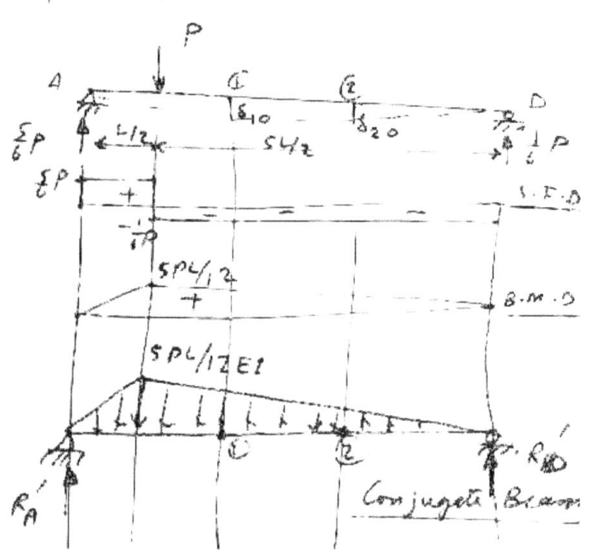

$$R_D = \frac{0.243\,PL^2}{EI}$$

$$M_1 = 0.243\frac{PL^2}{EI}(2L)$$

$$-\frac{1}{2}\left(\frac{2}{2.5}\right)\left(\frac{5PL}{12EI}\right)(2L)\left(\frac{1}{3}\right)(2L)$$

$$= 0.264\frac{PL^3}{EI}$$

$$\therefore\ \delta_{10} = -\frac{0.264\,PL^3}{EI}\ (down)$$

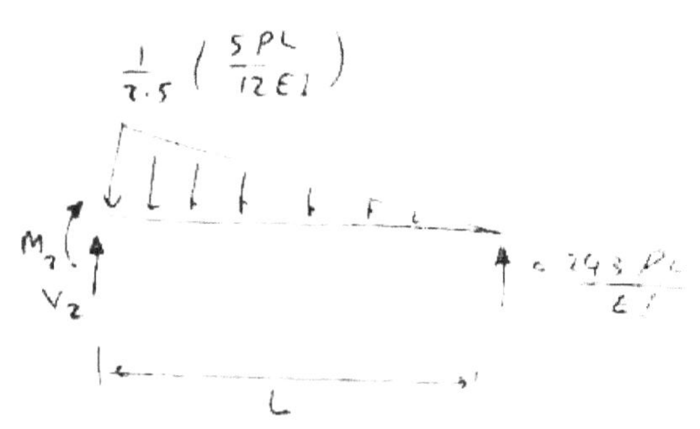

$$M_2 = 0.243\frac{PL^2}{EI}(L)$$

$$-\frac{1}{2}\left(\frac{1}{2.5}\right)\left(\frac{5PL}{12EI}\right)(L)\left(\frac{L}{3}\right)$$

$$= \frac{0.215\,PL^3}{EI}$$

$$\delta_{20} = -\frac{0.215\,PL^3}{EI}\ (down)$$

Step 2: Calculate f_{11} and f_{21} from second structure (1):

For the conjugate beam:

$+\circlearrowleft\ \Sigma M_A = 0.$

$$-R_B(3L) + \frac{1}{2}\left(\frac{2L}{3EI}\right)(2L)\left(L+\frac{2L}{3}\right)$$

$$+\frac{1}{2}\left(\frac{2L}{3EI}\right)(L)\left(\frac{2}{3}L\right) = 0$$

$$R_D = \frac{0.444\,L^2}{EI}$$

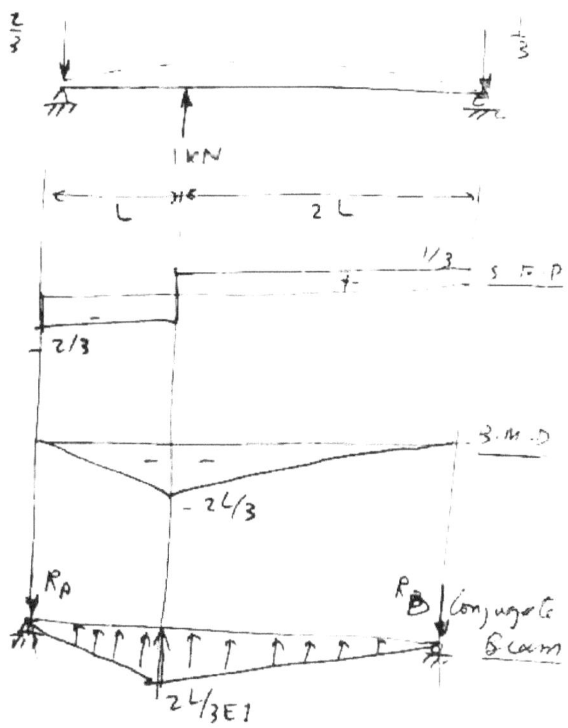

$$M_1 = -\frac{0.444 L^2}{EI}(2L) + \frac{1}{2}\left(\frac{2L}{3EI}\right)(2L)\left(\frac{1}{3}\right)(2L)$$

$$= -\frac{0.444 L^3}{EI}$$

$$\therefore f_{11} = +\frac{0.444 L^3}{EI} \quad (up)$$

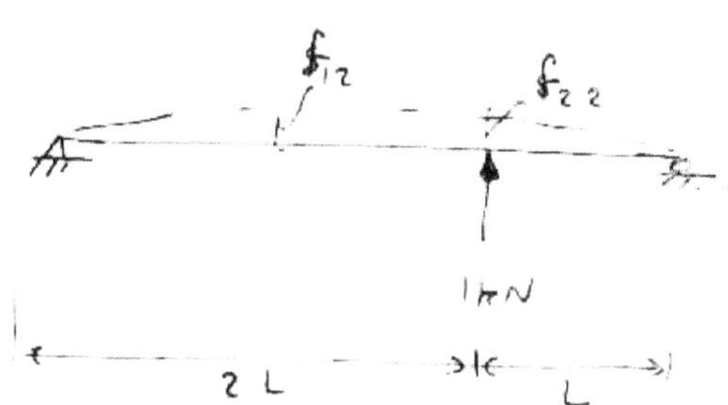

$$M_2 = -\frac{0.444 L^2}{EI}(L) + \frac{1}{2}\left(\frac{L}{3EI}\right)(L)\left(\frac{1}{3}L\right)$$

$$= -\frac{0.388 L^3}{EI}$$

$$\therefore f_{21} = +\frac{0.388 L^3}{EI} \quad (up)$$

Step 3 : Calculate f_{12} and f_{22} from second structure (2).

It is clear that

$$f_{12} = f_{21} = +\frac{0.388 L^3}{EI}$$

$$f_{22} = f_{11} = +\frac{0.444 L^3}{EI}$$

Step 4 : Equations of Consistent Deformations

$$\delta_{10} + R_1 f_{11} + R_2 f_{12} = 0 \quad \text{——} \quad (1)$$

$$\delta_{20} + R_1 f_{21} + R_2 f_{22} = 0 \quad \text{——} \quad (2)$$

$$-\frac{0.264 PL^3}{EI} + R_1\left(\frac{0.444 L^3}{EI}\right) + R_2\left(\frac{0.388 L^3}{EI}\right) = 0 \quad \text{——} \quad (1)$$

$$-\frac{0.215 PL^3}{EI} + R_1\left(\frac{0.388 L^3}{EI}\right) + R_2\left(\frac{0.444 L^3}{EI}\right) = 0 \quad \text{——} \quad (2)$$

Solve equations (1) and (2) simultaneously for R_1 and R_2 :

$$0.444 R_1 + 0.388 R_2 = 0.264 P \quad \text{——} \quad (1)$$

$$0.388 R_1 + 0.444 R_2 = 0.215 P \quad \text{——} \quad (2)$$

Multiply (1) by $-\dfrac{0.444}{0.388}$ and add it to (2)

$$-0.5081\,R_1 - 0.444\,R_2 = -0.3021\,P$$
$$0.388\,R_1 + 0.444\,R_2 = 0.215\,P$$

$$R_1 = 0.725\,P \uparrow$$

$$\therefore \quad R_2 = -0.149\,P \quad \Rightarrow \quad R_2 = 0.149\,P \downarrow$$

Step 5 : Equations of Equilibrium :

$\circlearrowright \ \Sigma M_A = 0 :$

$$R_D(3L) - 0.149\,P(2L) + 0.725\,P(L)$$
$$-P\left(\frac{L}{2}\right) = 0$$

$$R_D = 0.0241\,P \uparrow$$

$\uparrow \ \Sigma F_y = 0 :$

$$R_A - P + 0.725\,P - 0.149\,P + 0.0241\,P = 0$$

$$R_A = 0.40\,P \uparrow$$

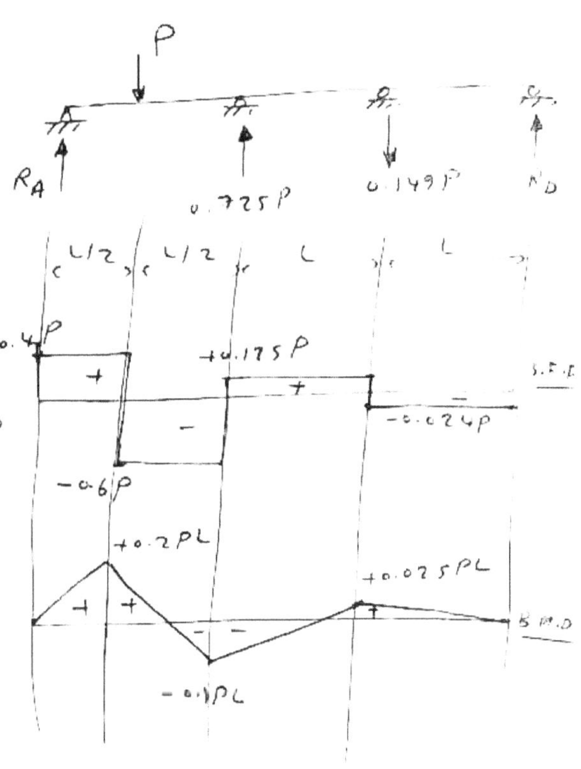

Example 11.8 :

Draw the bending moment diagram
for the frame shown in the figure.

Solution :

The frame is indeterminate to the
second degree (two redundants)

(Choose R_1 and R_2 as the redundants.

Equations of Consistent Deformations :

$$\delta_1 = 0 = \delta_{10} + R_1 f_{11} + R_2 f_{12} = 0 \quad ---- (1)$$

$$\delta_2 = 0 = \delta_{20} + R_1 f_{21} + R_2 f_{22} = 0 \quad ---- (2)$$

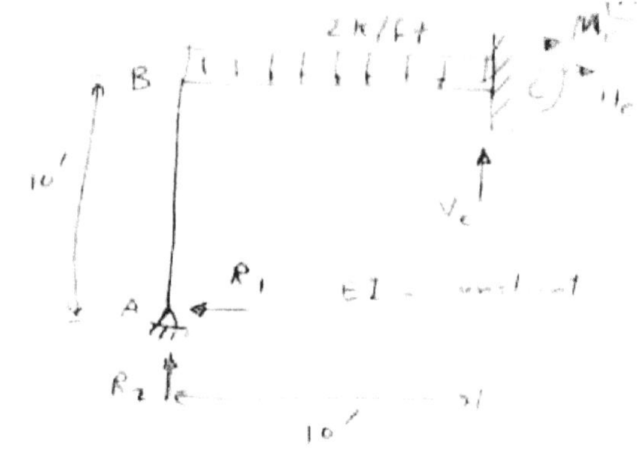

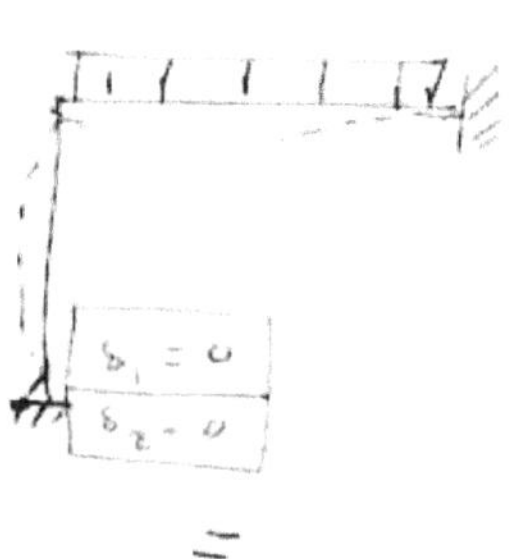

Use the method of virtual work.
In this case, the dummy structures
are the same as second structures
(1) & (2).

$$\delta_{10} = \int \frac{M m_1}{EI} dx$$

$$\delta_{20} = \int \frac{M m_2}{EI} dx$$

$$f_{11} = \int \frac{m_1^2}{EI} dx$$

$$f_{12} = \int \frac{m_1 m_2}{EI} dx$$

$$f_{21} = \int \frac{m_2 m_1}{EI} dx = f_{12}$$

$$f_{22} = \int \frac{m_2^2}{EI} dx$$

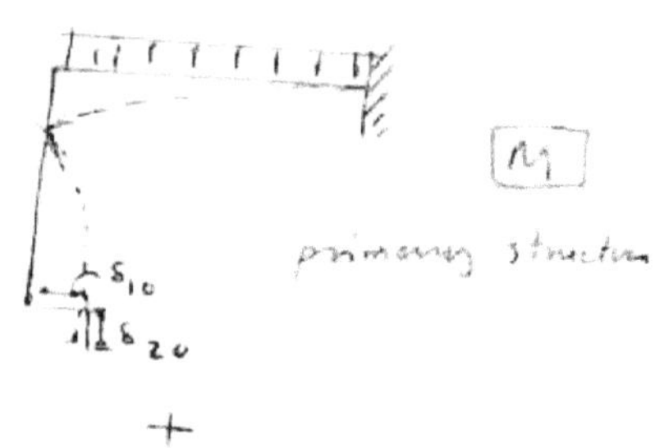

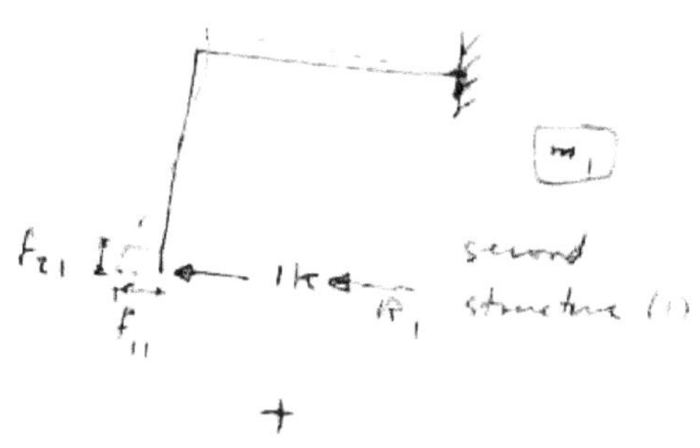

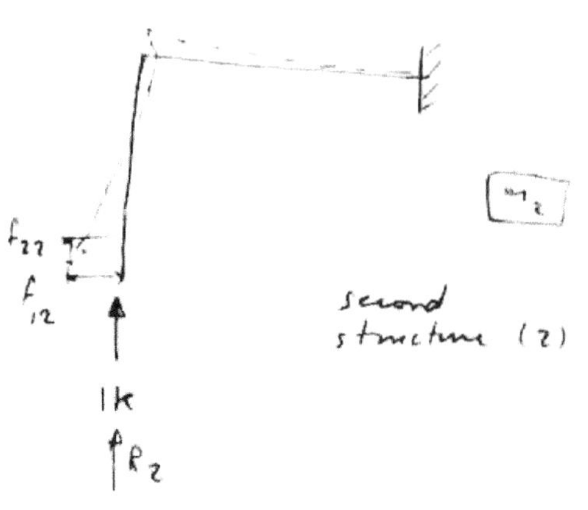

Member AB

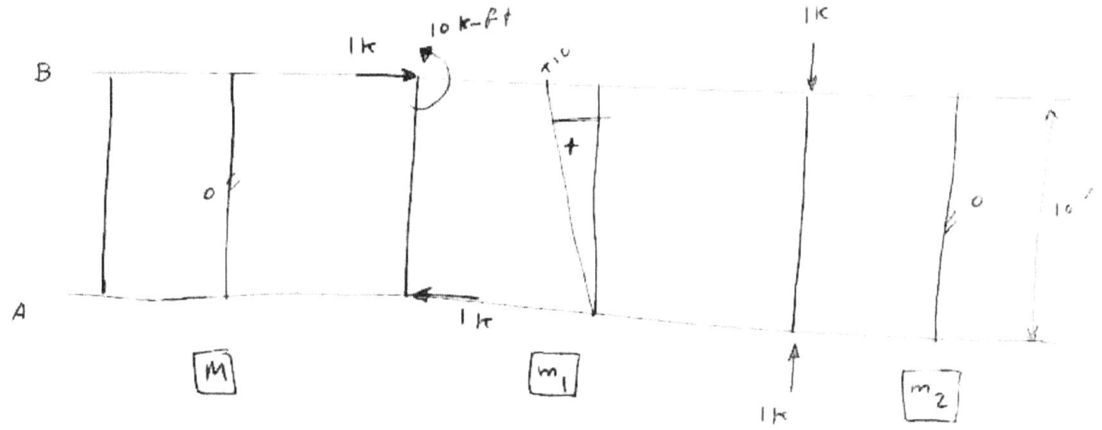

Member BC:

$$\delta_{10} = \frac{1}{EI} \int M m_1 \, dx$$

$$= 0 - \frac{1}{3}\left(\frac{100}{EI}\right)(10)(10) = -\frac{3333.333}{EI}$$

$$\delta_{20} = \frac{1}{EI} \int M m_2 \, dx$$

$$= \frac{1}{EI}\left[0 - \frac{1}{3}(100)(10)\left(\frac{3}{4}\right)(10) \right]$$

$$= -\frac{2500}{EI}$$

$$f_{11} = \frac{1}{EI} \int m_1^2 \, dx$$

$$= \frac{1}{EI}\left[\frac{1}{2}(10)(10)\left(\frac{2}{3}\right)(10) + (10)(10)(10) \right]$$

$$= +\frac{1333.333}{EI}$$

$$f_{12} = f_{21} = \frac{1}{EI} \int m_1 m_2 \, dx$$

$$= \frac{1}{EI}\left[0 + \frac{1}{2}(10)(10)(10) \right]$$

$$= +\frac{500}{EI}$$

$$f_{22} = \frac{1}{EI} \int m_2^2 \, dx$$

$$= \frac{1}{EI}\left[0 + \frac{1}{2}(10)(10)\left(\frac{2}{3}\right)(10) \right] = +\frac{333.333}{EI}$$

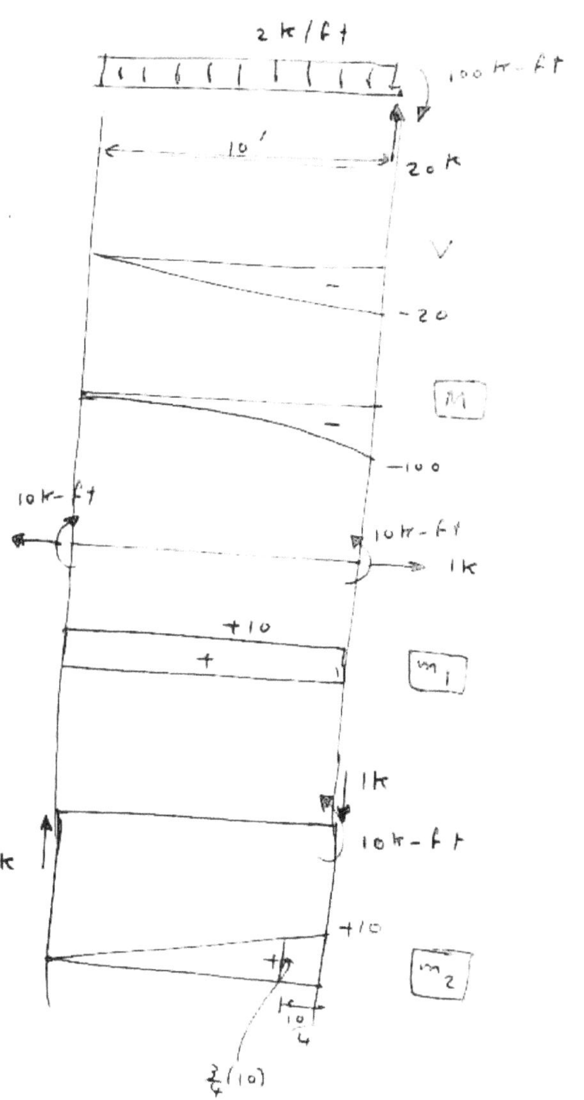

Substitute in the equations of consistent deformations:

$$\frac{1333.333}{EI} R_1 + \frac{500}{EI} R_2 = \frac{3333.333}{EI} \quad\text{------------ (1)}$$

$$\frac{500}{EI} R_1 + \frac{333.333}{EI} R_2 = \frac{2500}{EI} \quad\text{------------ (2)}$$

Solve eqs. (1) & (2) simultaneously for R_1 and R_2.

multiply (1) by $-\frac{333.333}{500}$ and add it to (2).

$$-888.888 R_1 - 333.333 R_2 = -2227.22$$
$$500 R_1 + 333.333 R_2 = 2500$$

$$\therefore R_1 = -0.714 \, k \implies R_1 = 0.714 \, k \to$$

$$\therefore R_2 = +8.57 \, k \uparrow$$

Equations of Equilibrium:

$\xrightarrow{+} \Sigma F_x = 0 : \quad H_c = 0.714 \, k \leftarrow$

$+\uparrow \Sigma F_y = 0 :$

$\quad V_c + 8.57 - 2(10) = 0$

$\quad V_c = 11.43 \, k \uparrow$

$+\circlearrowleft \Sigma M_c = 0 :$

$M_c + 2(10)(5) + 0.714(10) - 8.57(10) = 0$

$M_c = -21.44 \, k\text{-}ft \implies M_c = 21.44 \circlearrowright k\text{-}ft$

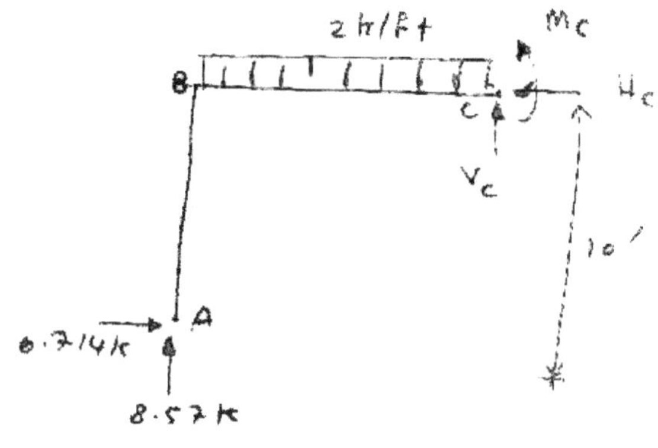

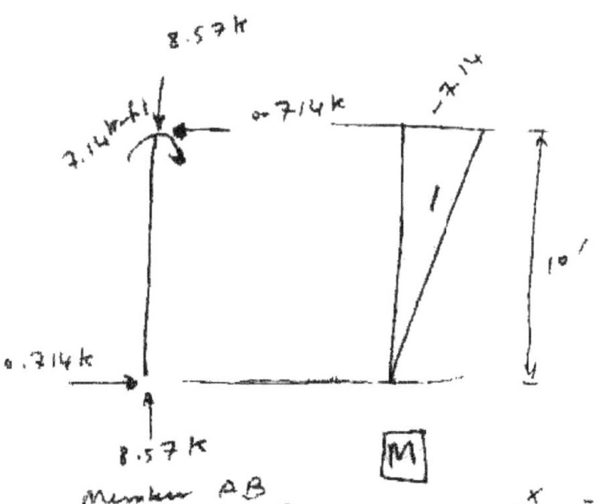

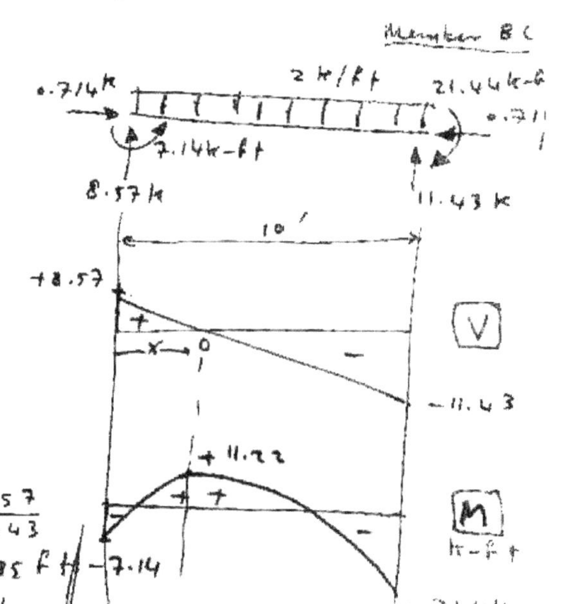

Member AB
Locate the point of zero shear: $\dfrac{x}{10-x} = \dfrac{8.57}{11.43}$

$11.43 x = 8.57(10) - 8.57 x \implies x = 4.285 \, ft$

$M_0 = -7.14 + \frac{1}{2}(8.57)(4.285) = +11.22 \, k\text{-}ft$

11.6 Structures with Internal Redundants.

* Use numbers to denote members.

Example 11.9: Calculate all member forces of the truss shown in the figure.

$$r = 3 \atop m = 6 \Big\} \quad r + m = 9$$

$$j = 4, \quad 2j = 8$$

$$\therefore \quad r + m > 2j$$

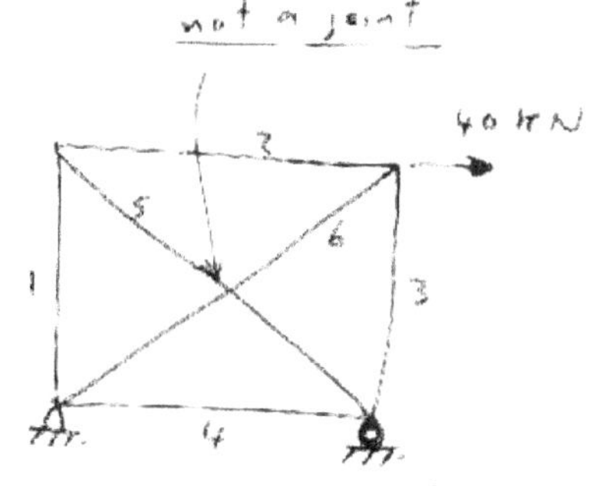

The truss is indeterminate to the first degree
   (one redundant).

But we can determine the three reactions

∴ The truss is _internally_ indeterminate

   Choose member 6 as the redundant (i.e. F_6)

Use the method of consistent deformations.

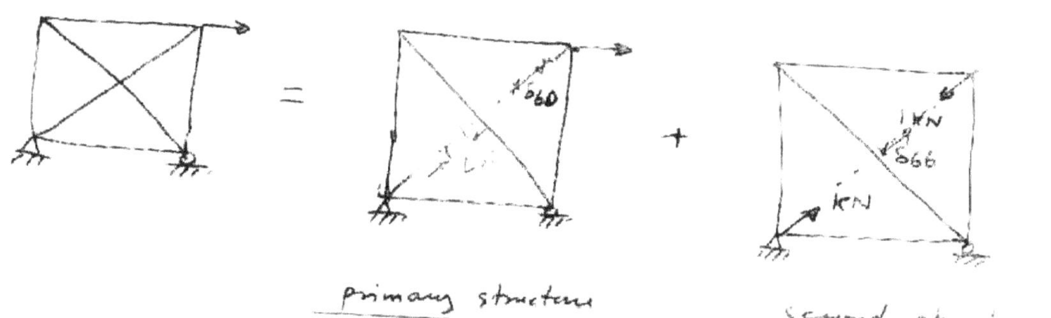

$$\delta_6 = 0 = \delta_{60} + F_6 \delta_{66} = 0.$$

Step 1 ; Calculate δ_{60} from the primary structure.

$$\delta_{60} = \sum \frac{N n L}{EA} = \frac{1}{EA} \sum N n L$$

Step 2 ; Calculate δ_{66} from the second structure ;

$$\delta_{66} = \sum \frac{n n L}{EA} = \frac{1}{EA} \sum n^2 L$$

Primary Structure :

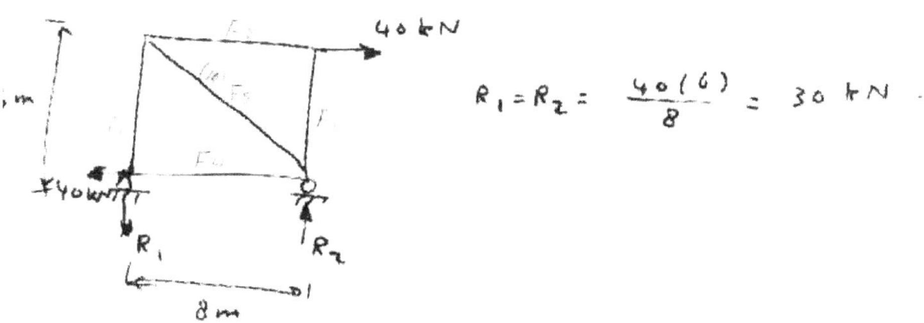

$$R_1 = R_2 = \frac{40(6)}{8} = 30 \text{ kN}.$$

$\bar{F}_1 = +30 \ kN$

$F_4 = +40 \ kN$

$+\uparrow \Sigma F_y = 0: \quad -30 - F_5 \left(\frac{6}{10}\right) = 0$

$F_5 = -50 \ kN$

$\xrightarrow{+} \Sigma F_x = 0 \quad F_2 + (-50)\left(\frac{8}{10}\right) = 0$

$F_2 = +40 \ kN$

$F_3 = 0$

Second Structure:

$r_2 = 0$

$\xrightarrow{+} \Sigma F_x = 0: \quad r_3 = 0$

$+\uparrow \Sigma F_y = 0. \quad r_1 = 0$

$\xrightarrow{+} \Sigma F_x = 0: \quad f_4 + 1\left(\frac{8}{10}\right) = 0 \implies f_4 = -0.8 \ kN$

$+\uparrow \Sigma F_y = 0: \quad f_1 + 1\left(\frac{6}{10}\right) = 0 \implies f_1 = -0.6$

$+\uparrow \Sigma F_y = 0: \quad -F_5 \left(\frac{6}{10}\right) - (-0.6) = 0$

$f_5 = +1 \ kN$

$\xrightarrow{+} \Sigma F_x = 0. \quad f_2 + (1)\left(\frac{8}{10}\right) = 0 \implies f_2 = -0.8 \ kN$

$+\uparrow \Sigma F_y = 0: \quad -1\left(\frac{6}{10}\right) - f_3 = 0$

$f_3 = -0.6 \ kN$

Member	N (kN)	n (kN)	L (m)	NnL	$n^2 L$
1	+30	−0.6	6	−108	2.16
2	+40	−0.8	8	−256	5.12
3	0	−0.6	6	0	2.16
4	+40	−0.8	8	−256	5.12
5	−50	+1.0	10	−500	10.0
6	0	+1.0	10	0	10.0
				−1120	34.56

$$\delta_{6O} = \frac{1}{EA} \sum N_n L = \frac{-1120}{EA}$$

$$\delta_{66} = \frac{1}{EA} \sum n^2 L = \frac{34.56}{EA}$$

Step 3 : Equation of Consistent Deformation :

$$\delta_{6O} + F_6 \, S_{66} = 0$$

$$\frac{-1120}{EA} + F_6 \left(\frac{34.56}{EA} \right) = 0 \implies F_6 = + \underline{32.41 \, kN \, (T)}$$

Step 4 : Determine the other member forces using the method of joint

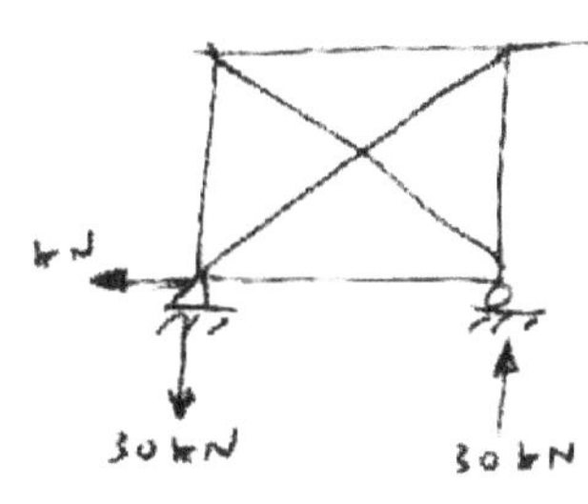

$$\rightarrow \sum F_x = 0 : \quad F_4 + 32.41\left(\tfrac{8}{10}\right) - 40 = 0$$

$$F_4 = + \underline{14.07} \; kN \; (T).$$

$$+\uparrow \sum F_y = 0 : \quad F_1 + (32.41)\left(\tfrac{6}{10}\right) - 30 = 0$$

$$F_1 = +10.55 \; (T).$$

$$+\uparrow \sum F_y = 0 : \quad -F_5\left(\tfrac{6}{10}\right) - 10.55 = 0$$

$$F_5 = -17.59 \; kN \quad (C).$$

$$\xrightarrow{+} \sum F_x = 0 : \quad (-17.59)\left(\tfrac{8}{10}\right) + F_2 = 0$$

$$F_2 = +14.07 \; (T).$$

$$+\uparrow \sum F_y = 0 : \quad -32.41\left(\tfrac{6}{10}\right) - F_3 = 0$$

$$F_3 = -19.45 \; kN \; (C)$$

Example A1:

Find the reaction at B and the bar force in member BF of the truss shown in the figure. The cross-sectional areas of the members in cm² are shown on the figure. E = constant.

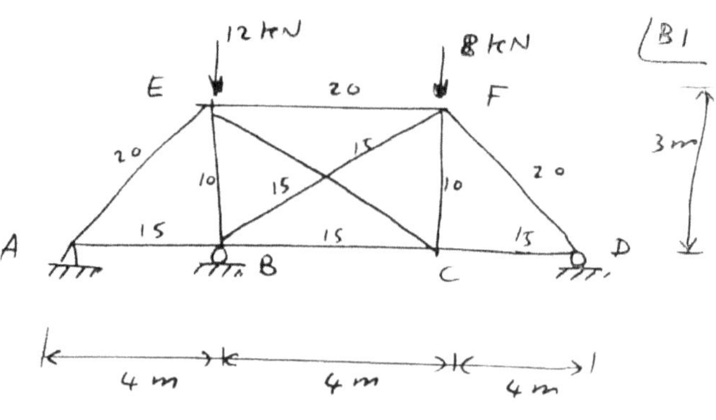

Solution:

$$r = 4$$
$$m = 10 \quad\Big\}\quad m + r = 14 \quad\Big\}\quad 14 - 12 = 2.$$
$$j = 6 \implies 2j = 12$$

∴ The truss is indeterminate to the second degree.
(we have **two** redundants).

Choose R_B and F_5 as the redundants.

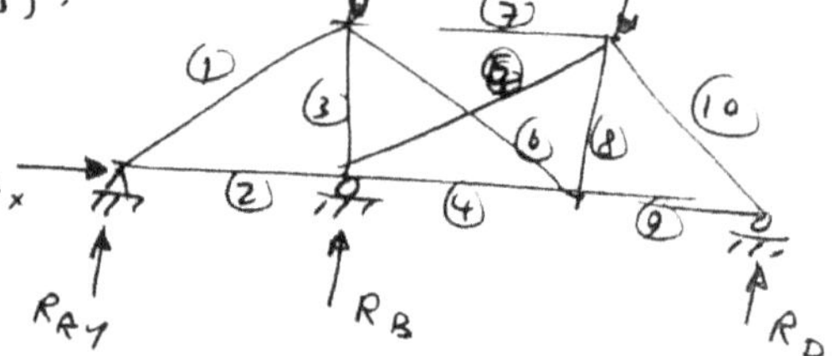

Equations of Consistent Deformations:

$$\delta_B = 0 = \delta_{Bo} + R_B \delta_{BB} + F_5 \delta_{B5} = 0 \quad \text{---(1)}$$
$$\delta_5 = 0 = \delta_{5o} + R_B \delta_{5B} + F_5 \delta_{55} = 0 \quad \text{---(2)}$$

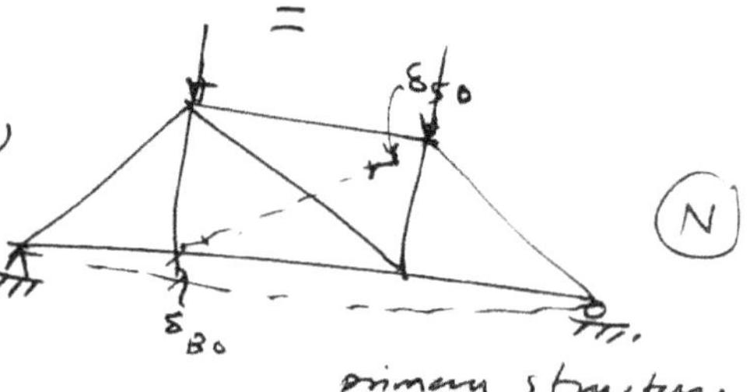

Step 1: Calculate δ_{Bo} & δ_{5o} from the primary structure:

$$\delta_{Bo} = \sum \frac{N n_1 L}{EA} = \frac{1}{E} \sum \frac{N n_1 L}{A}$$

$$\delta_{5o} = \sum \frac{N n_2 L}{EA} = \frac{1}{E} \sum \frac{N n_2 L}{A}$$

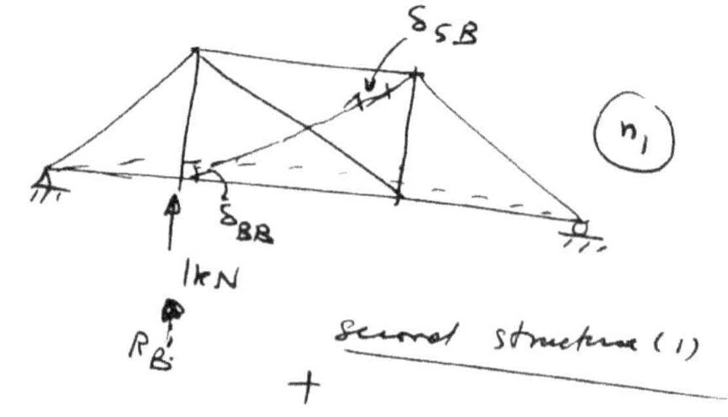

Step 2: Calculate δ_{BB} & δ_{5B} from the second structure (1):

$$\delta_{BB} = \sum \frac{n_1 n_1 L}{EA} = \frac{1}{E} \sum \frac{n_1^2 L}{A}$$

$$\delta_{5B} = \sum \frac{n_1 n_2 L}{EA} = \frac{1}{E} \sum \frac{n_1 n_2 L}{A}$$

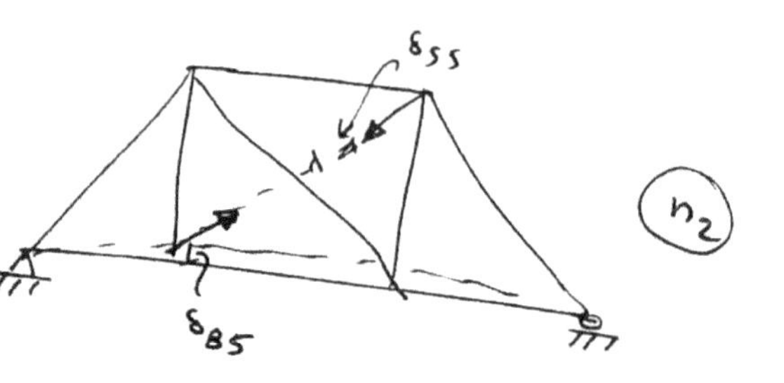

Step 3 : Calculate δ_{BS} & δ_{SS} from the second structure (2) :

$$\delta_{BS} = \sum \frac{n_2 n_1 L}{EA} = \frac{1}{E} \sum \frac{n_2 n_1 L}{A}$$

$$\delta_{SS} = \sum \frac{n_2 n_2 L}{EA} = \frac{1}{E} \sum \frac{n_2^2 L}{A}$$

Note that $\boxed{\delta_{SB} = \delta_{BS}}$ by Maxwell's Law of Reciprocal Deflections.

Primary Structure :

$R_{Ax} = 0$

$R_{Ay} = 12\left(\frac{8}{12}\right) + 8\left(\frac{4}{12}\right) = 10.67 \text{ kN} \uparrow$

$R_D = 12\left(\frac{4}{12}\right) + 8\left(\frac{8}{12}\right) = 9.33 \text{ kN} \uparrow$

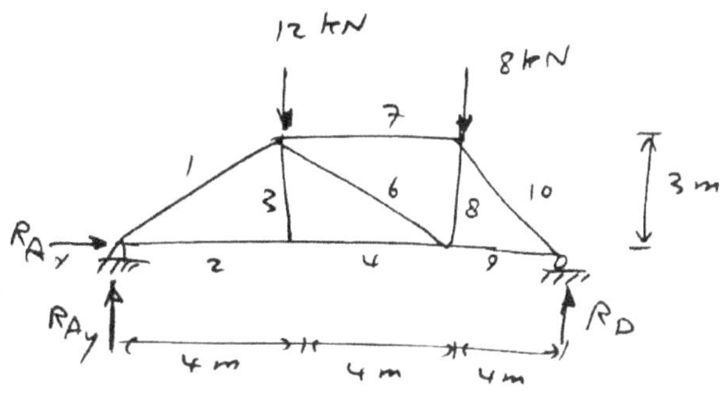

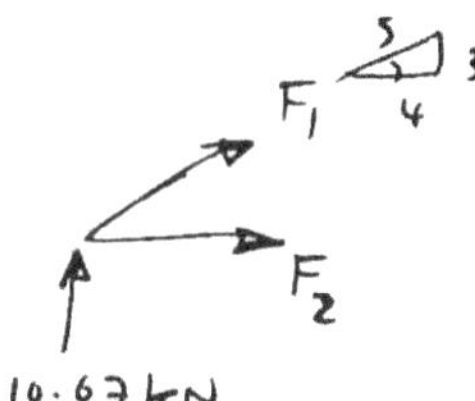

$F_1\left(\frac{3}{5}\right) + 10.67 = 0 \implies F_1 = -17.78 \text{ kN}$

$F_2 + (-17.78)\left(\frac{4}{5}\right) = 0 \implies F_2 = +14.23 \text{ kN}$

$F_3 = 0$

$F_4 = +14.23 \text{ kN}$

$-12 - (-17.78)\left(\frac{3}{5}\right) - F_6\left(\frac{3}{5}\right) = 0$

$F_6 = -2.22 \text{ kN}$

$F_7 + (-2.22)\left(\frac{4}{5}\right) - (-17.78)\left(\frac{4}{5}\right) = 0$

$F_7 = -12.45 \text{ kN}$

$F_{10}\left(\frac{4}{5}\right) + 12.45 = 0 \implies F_{10} = -15.56 \text{ kN}$

$-F_8 - 8 - (-15.56)\left(\frac{3}{5}\right) = 0 \implies F_8 = +1.34 \text{ kN}$

$-F_9 + 15.56\left(\frac{4}{5}\right) = 0 \implies F_9 = +12.45 \text{ kN}$

$\sum F_x = 12.45 - 14.23 + 2.22\left(\frac{4}{5}\right) = 0$

$\sum F_y = 1.34 - 2.22\left(\frac{3}{5}\right) = 0$

Second Structure (1):

$$R_A = 1\left(\frac{8}{12}\right) = 0.667 \text{ kN} \downarrow$$

$$R_D = 1\left(\frac{4}{12}\right) = 0.333 \text{ kN} \downarrow$$

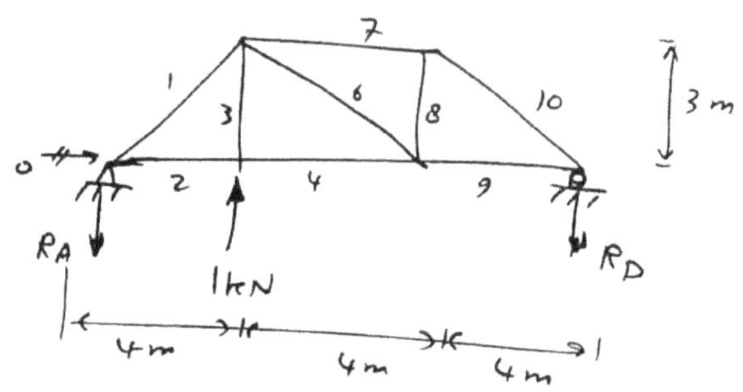

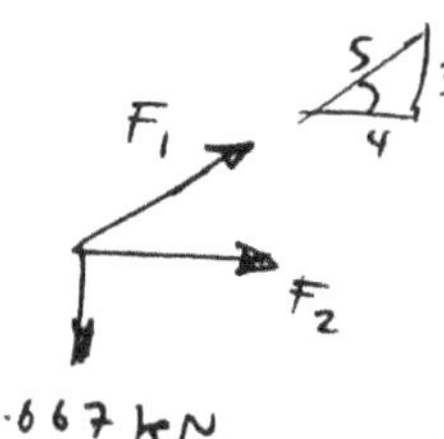

$$F_1\left(\frac{3}{5}\right) - 0.667 = 0 \implies F_1 = +1.112 \text{ kN}$$

$$F_2 + (1.112)\left(\frac{4}{5}\right) = 0 \implies F_2 = -0.889 \text{ kN}$$

$$F_3 = -1 \text{ kN}$$

$$F_4 = -0.889 \text{ kN}$$

$$-F_6\left(\frac{3}{5}\right) + 1 - (1.112)\left(\frac{3}{5}\right) = 0$$

$$F_6 = +0.555 \text{ kN}$$

$$F_7 + (0.555)\left(\frac{4}{5}\right) - 1.112\left(\frac{4}{5}\right) = 0$$

$$F_7 = +0.446 \text{ kN}$$

$$F_{10}\left(\frac{4}{5}\right) - 0.446 = 0 \implies F_{10} = +0.557 \text{ kN}$$

$$-F_8 - (+0.557)\left(\frac{3}{5}\right) = 0 \implies F_8 = -0.334 \text{ kN}$$

$$-F_9 - 0.557\left(\frac{4}{5}\right) = 0 \implies F_9 = -0.446 \text{ kN}$$

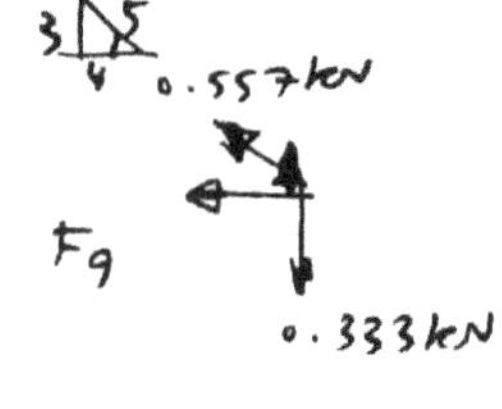

$$\Sigma F_x = -0.446 + 0.889 - 0.555\left(\frac{4}{5}\right) = 0 ✓$$

$$\Sigma F_y = -0.334 + 0.555\left(\frac{5}{6}\right) = 0 ✓$$

Second Structure (2):

$$R_D = 0$$

$$R_{Ay} = R_{Ax} = 0$$

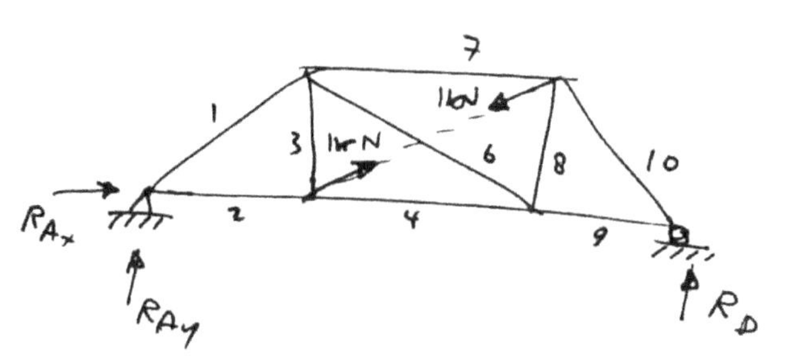

$$F_1 = 0$$

$$F_2 = 0$$

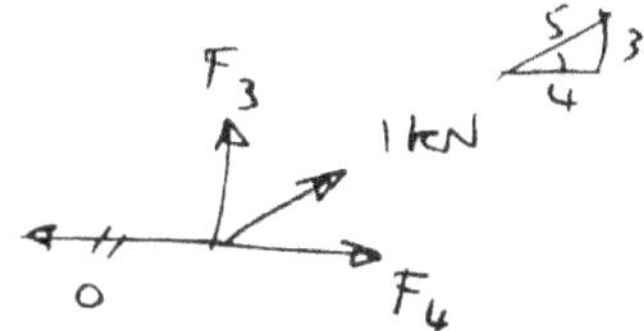

$$+1\left(\tfrac{4}{5}\right) + F_4 = 0 \implies F_4 = -0.8 \text{ kN}$$

$$+1\left(\tfrac{3}{5}\right) + F_3 = 0 \implies F_3 = -0.6 \text{ kN}$$

$$-F_6\left(\tfrac{3}{5}\right) + 0.6 = 0 \implies F_6 = +1.0 \text{ kN}$$

$$F_7 + 1\left(\tfrac{4}{5}\right) = 0 \implies F_7 = -0.8 \text{ kN}$$

$$F_{10}\left(\tfrac{4}{3}\right) - 1\left(\tfrac{4}{5}\right) + 0.8 = 0 \implies F_{10} = 0$$

$$-F_8 - (0)\left(\tfrac{3}{5}\right) - 1\left(\tfrac{3}{5}\right) = 0$$

$$F_8 = -0.6 \text{ kN}$$

$$F_9 = 0$$

Member	L (m)	N (kN)	n_1 (kN)	n_2 (kN)	A (cm²)	$\dfrac{Nn_1L}{A}$	$\dfrac{Nn_2L}{A}$	$\dfrac{n_1^2L}{A}$	$\dfrac{n_2^2L}{A}$	$\dfrac{n_1n_2L}{A}$
1	5	−17.78	+1.112	0	20	−4.943	0	0.3091	0	0
2	4	+14.23	−0.889	0	15	−3.373	0	0.2108	0	0
3	3	0	−1.0	−0.6	10	0	0	0.3	0.108	0.18
4	4	+14.23	−0.889	−0.8	15	−3.373	−3.036	0.2108	0.1707	0.1897
5	5	0	0	+1.0	15	0	0	0	0.3333	0
6	5	−2.22	+0.556	+1.0	15	−0.411	−0.740	0.1030	0.3333	0.1853
7	4	−12.45	+0.446	−0.8	20	−1.111	+1.992	0.0398	0.128	−0.0714
8	3	+1.34	−0.334	−0.6	10	−0.134	−0.2412	0.0335	0.108	0.0601
9	4	+12.45	−0.446	0	15	−1.481	0	0.0530	0	0
10	5	−15.56	+0.557	0	20	−2.167	0	0.0776	0	0
						−16.993	−2.0252	+1.3376	+1.1813	+0.5437

Step 4 : Equations of Consistent Deformation :

$$\delta_{Bo} + R_B f_{BB} + F_S f_{BS} = 0 \quad \text{——— (1)}$$

$$\delta_{So} + R_B f_{SB} + F_S f_{SS} = 0 \quad \text{——— (2)}$$

$$-\frac{16.993}{E} + R_B \left(\frac{1.3376}{E}\right) + F_S \left(\frac{0.5437}{E}\right) = 0 \quad \text{——— (1)}$$

$$-\frac{2.0252}{E} + R_B \left(\frac{0.5437}{E}\right) + F_S \left(\frac{1.1813}{E}\right) = 0 \quad \text{——— (2)}$$

$$\begin{cases} -16.993 + 1.3376\, R_B + 0.5437\, F_S = 0 \quad \text{——— (1)} \\ -2.0252 + 0.5437\, R_B + 1.1813\, F_S = 0 \quad \text{——— (2)} \end{cases}$$

Solve eqs. (1) & (2) simultaneously :

multiply (1) by $\quad -\dfrac{0.5437}{1.3376} \quad$:

$$6.9072 - 0.5437\, R_B - 0.2210\, F_S = 0$$

$$-2.0252 + 0.5437\, R_B + 1.183\, F_S = 0$$

$$\Rightarrow \quad F_S = -5.07 \text{ kN} \quad (c).$$

$$R_B = +14.77 \text{ kN} \uparrow .$$

Structures with Three Redundants:

Consider the continuous beam shown in the figure.

The beam is indeterminate to the $\underline{\text{third}}$ degree.

Choose R_A, R_B, R_C as the redundants.

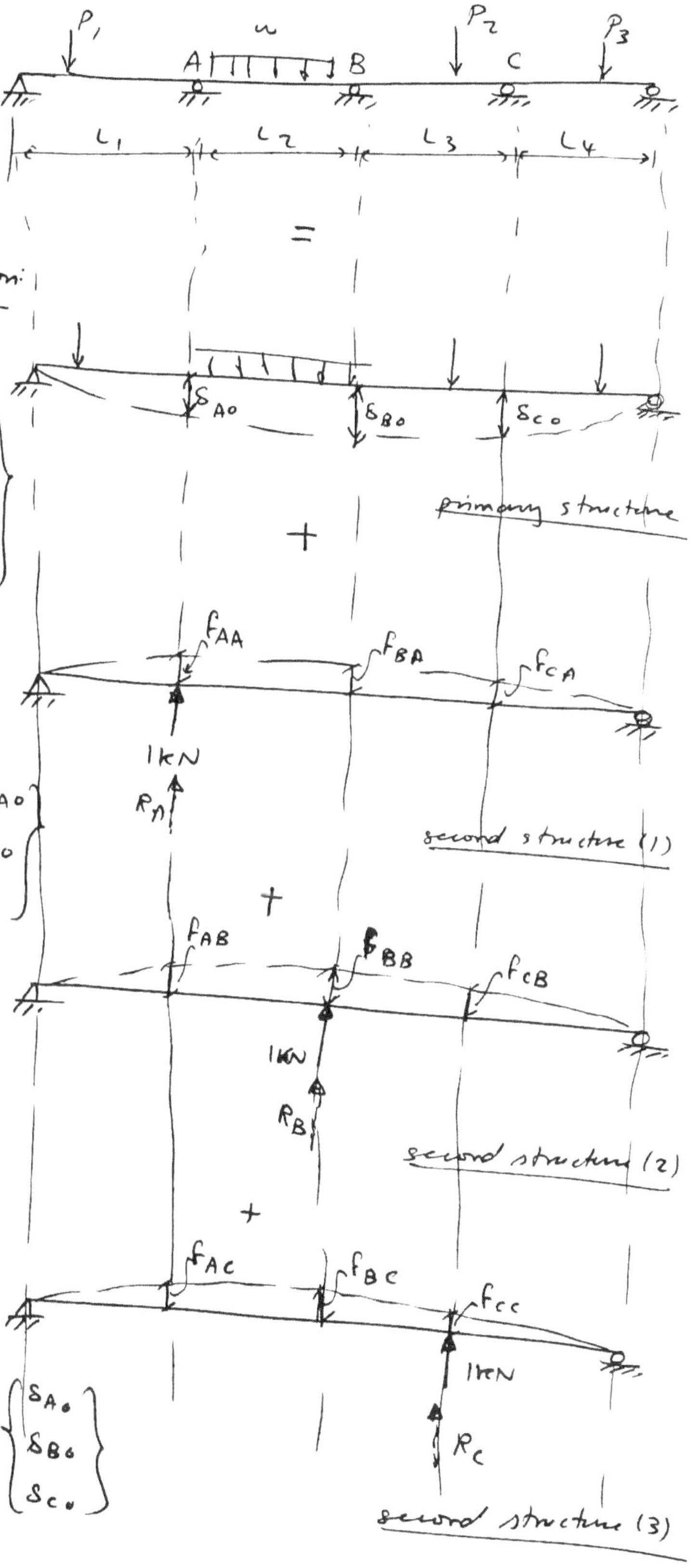

Equations of Consistent Deformation:

$$\delta_A = 0 = \delta_{AO} + R_A f_{AA} + R_B f_{AB} + R_C f_{AC}$$

$$\delta_B = 0 = \delta_{BO} + R_A f_{BA} + R_B f_{BB} + R_C f_{BC}$$

$$\delta_C = 0 = \delta_{CO} + R_A f_{CA} + R_B f_{CB} + R_C f_{CC}$$

Rewrite the equations in matrix form:

$$\begin{bmatrix} f_{AA} & f_{AB} & f_{AC} \\ f_{BA} & f_{BB} & f_{BC} \\ f_{CA} & f_{CB} & f_{CC} \end{bmatrix} \begin{Bmatrix} R_A \\ R_B \\ R_C \end{Bmatrix} = - \begin{Bmatrix} \delta_{AO} \\ \delta_{BO} \\ \delta_{CO} \end{Bmatrix}$$

$\nearrow$ flexibility matrix

But, using Maxwell's law of reciprocal deflections:

$$f_{BA} = f_{AB}$$

$$f_{CA} = f_{AC}$$

$$f_{CB} = f_{BC}$$

$$\begin{bmatrix} f_{AA} & f_{AB} & f_{AC} \\ f_{AB} & f_{BB} & f_{BC} \\ f_{AC} & f_{BC} & f_{CC} \end{bmatrix} \begin{Bmatrix} R_A \\ R_B \\ R_C \end{Bmatrix} = - \begin{Bmatrix} \delta_{AO} \\ \delta_{BO} \\ \delta_{CO} \end{Bmatrix}$$

$\nearrow$

The flexibility matrix is $\underline{\text{symmetric}}$

Example A2 :

Determine the reactions at support D of the frame shown in the figure.

Solution :

The frame is indeterminate to the third degree (three redundants).

Choose D_x, D_y, M_D as the redundants.

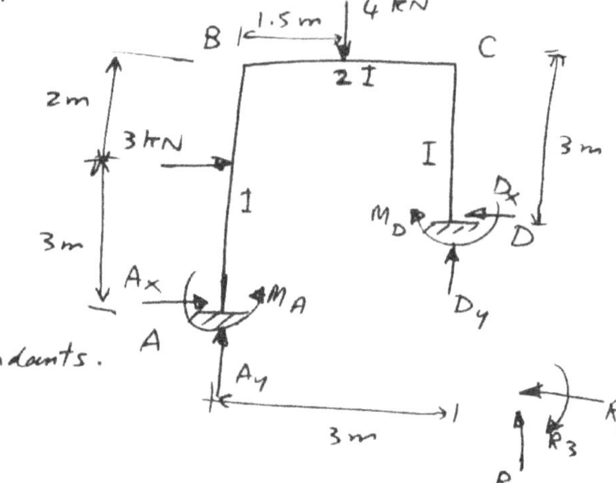

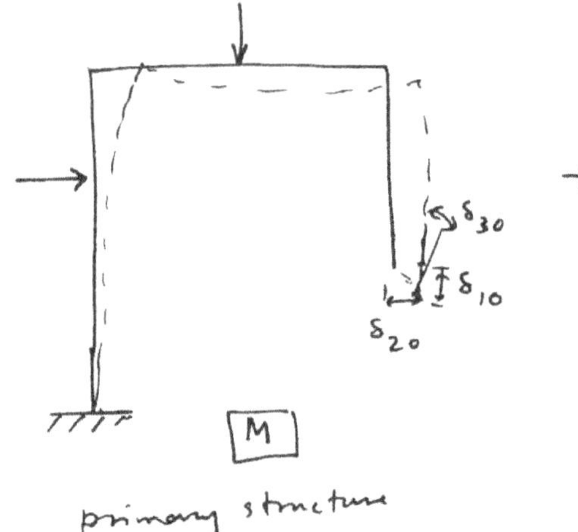

primary structure

$+$

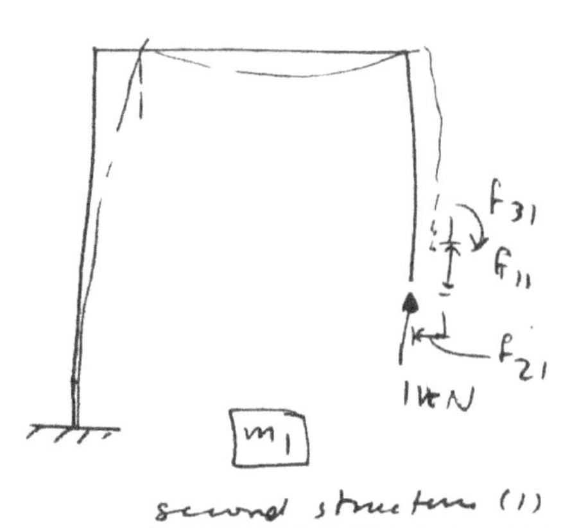

second structure (1)

$+$

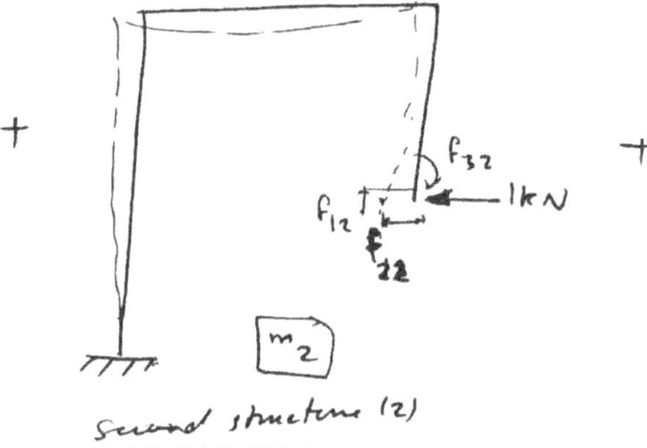

second structure (2)

$+$

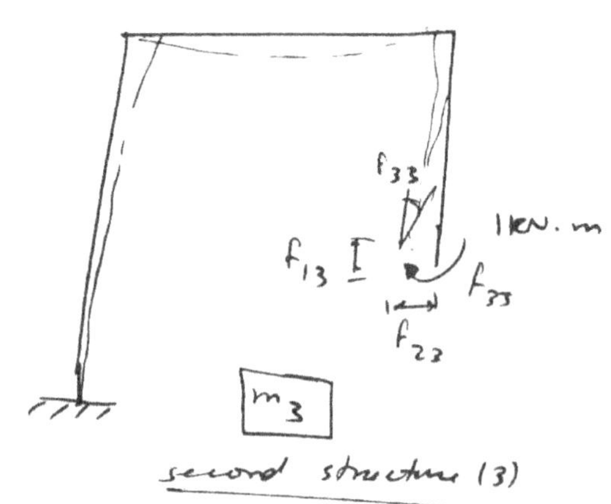

second structure (3)

Equations of Consistent Deformation:

$$\begin{bmatrix} f_{11} & f_{12} & f_{13} \\ f_{12} & f_{22} & f_{23} \\ f_{13} & f_{23} & f_{33} \end{bmatrix} \begin{Bmatrix} R_1 \\ R_2 \\ R_3 \end{Bmatrix} = - \begin{Bmatrix} \delta_{10} \\ \delta_{20} \\ \delta_{30} \end{Bmatrix}$$

Step 1: Calculate $\delta_{10}, \delta_{20}, \delta_{30}$ from the primary structure:

$$\delta_{10} = \int \frac{M m_1}{EI} dx \quad , \quad \delta_{20} = \int \frac{M m_2}{EI} dx \quad , \quad \delta_{30} = \int \frac{M m_3}{EI} dx$$

Step 2: Calculate f_{11}, f_{21}, f_{31} from the second structure (1):

$$f_{11} = \int \frac{m_1 m_1}{EI} dx = \int \frac{m_1^2}{EI} dx \quad , \quad f_{21} = \int \frac{m_1 m_2}{EI} dx \quad , \quad f_{31} = \int \frac{m_1 m_3}{EI} dx$$

Step 3: Calculate f_{12}, f_{22}, f_{32} from the second structure (2):

$$f_{12} = \int \frac{m_2 m_1}{EI} dx = f_{21} \quad , \quad f_{22} = \int \frac{m_2^2}{EI} dx \quad , \quad f_{23} = \int \frac{m_3 m_2}{EI} dx$$

Step 4: Calculate f_{13}, f_{23}, f_{33} from the second structure (3):

$$f_{13} = \int \frac{m_3 m_1}{EI} dx \quad , \quad f_{23} = \int \frac{m_3 m_2}{EI} dx = f_{32} \quad , \quad f_{33} = \int \frac{m_3^2}{EI} dx$$

Calculate the reactions for each structure:

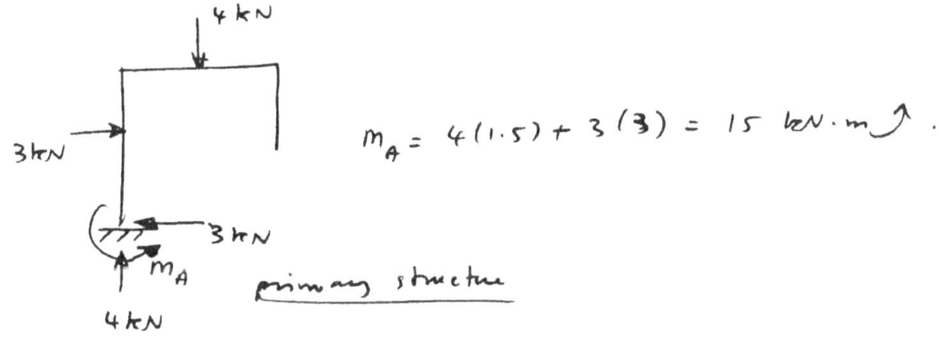

$$m_A = 4(1.5) + 3(3) = 15 \text{ kN·m} \uparrow$$

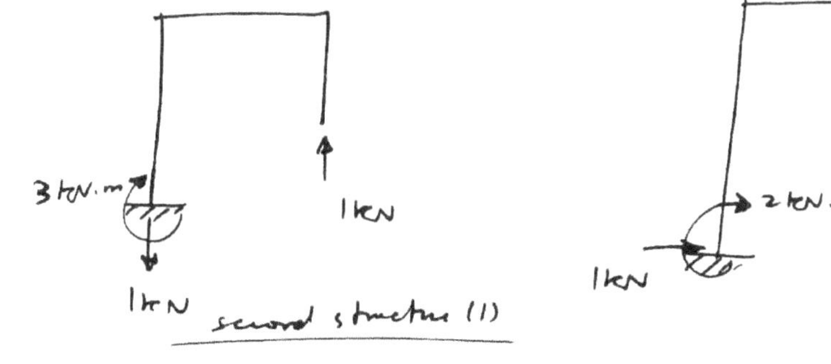

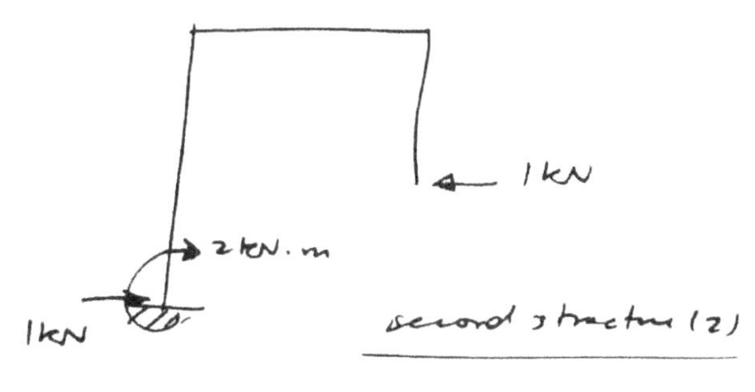

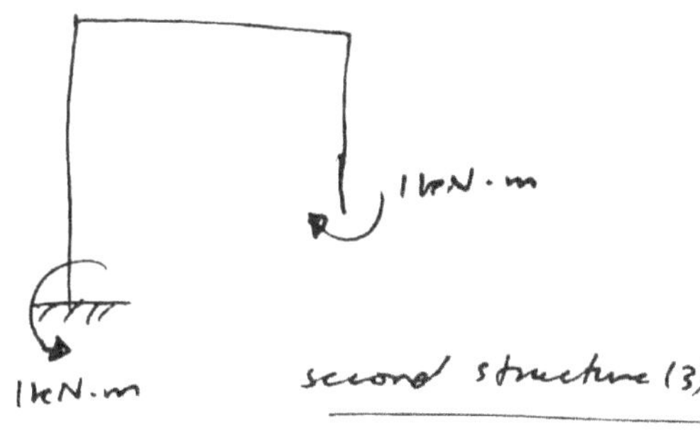

Member AB :

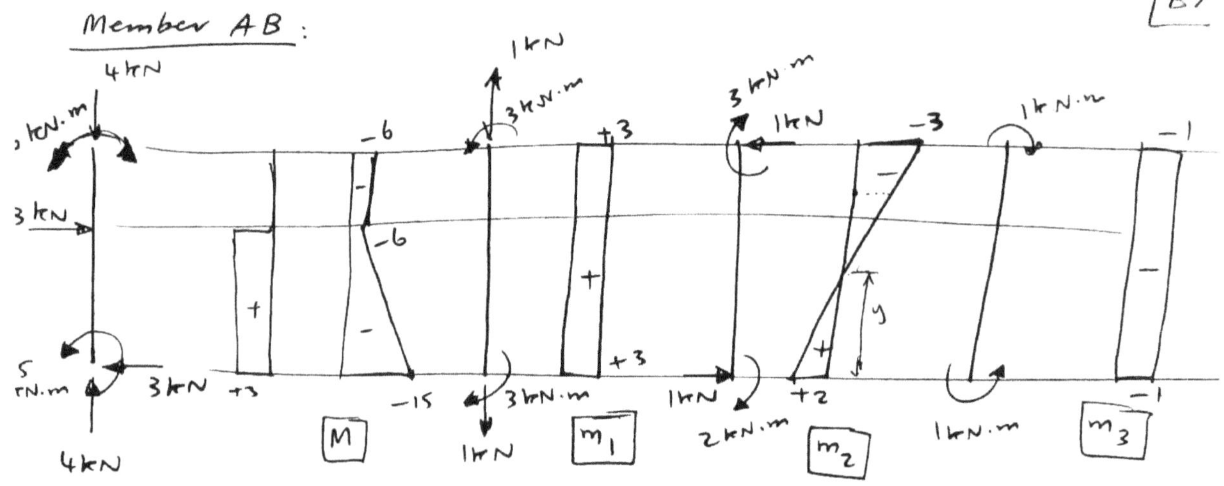

Member BC :

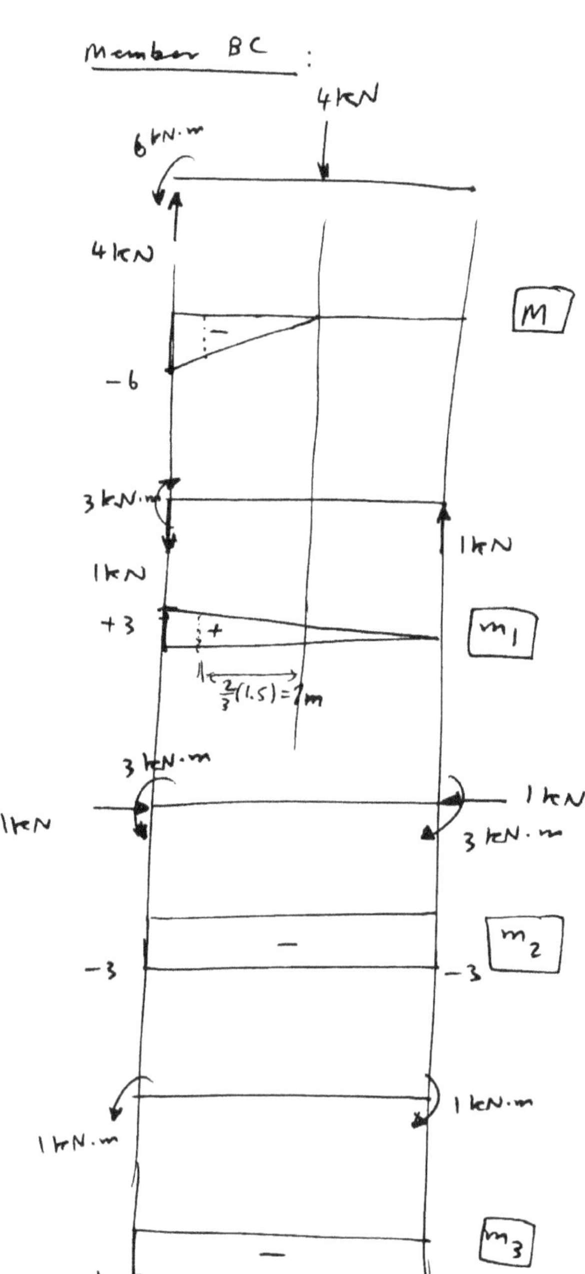

$$\frac{y}{5-y} = \frac{2}{3}$$

$$3y = 10 - 2y$$

$$5y = 10 \implies y = 2\,m.$$

Note

Member	(I)
AB	I
BC	2I
CD	I

Member CD :

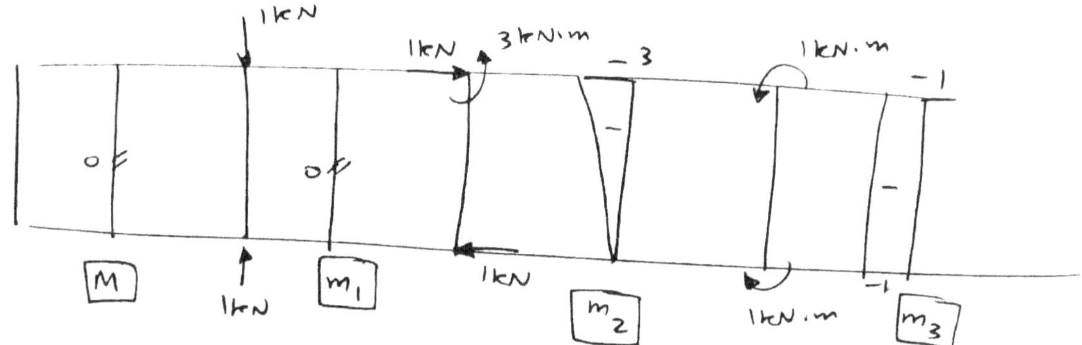

$$\delta_{10} = \frac{1}{EI} \int M m_1 \, dx = \frac{1}{EI}\left(-6(2) - \left[\frac{6+15}{2}\right](3) \right)(3) + \frac{1}{2EI}\left[-\frac{1}{2}(6)(1.5)(2.5) \right]$$

$$+ 0$$

$$= -136.125/EI$$

$$\delta_{20} = \frac{1}{EI} \int M m_2 \, dx = \frac{1}{EI}\left[(-6)(5)(\bar{0}.5) - \frac{1}{2}(9)(3)(+1) \right] + \frac{1}{2EI}\left[-\frac{1}{2}(6)(1.5)(-3) \right]$$

$$+ 0$$

$$= +8.25/EI$$

$$\delta_{30} = \frac{1}{EI} \int M m_3 \, dx = \frac{1}{EI}\left[-6(2) - \frac{1}{2}(15+6)(3) \right](-1) + \frac{1}{2EI}\left[-\frac{1}{2}(6)(1.5)(-1) \right]$$

$$+ 0$$

$$= +45.75/EI$$

$$f_{11} = \frac{1}{EI} \int m_1^2 \, dx = \frac{1}{EI}\left[3(5)(3) \right] + \frac{1}{2EI}\left[\frac{1}{2}(3)(3)(2) \right] + 0$$

$$= +49.5/EI$$

$$f_{12} = f_{21} = \frac{1}{EI} \int m_1 m_2 \, dx = \frac{1}{EI}\left[-\frac{1}{2}(3)(3) + \frac{1}{2}(2)(2) \right](3) + \frac{1}{2EI}\left[\frac{1}{2}(3)(3)(-3) \right]$$

$$+ 0$$

$$= -14.25/EI$$

$$f_{13} = f_{31} = \frac{1}{EI} \int m_1 m_3 \, dx = \frac{1}{EI}\left[(3)(5)(-1) \right] + \frac{1}{2EI}\left[\frac{1}{2}(3)(3)(-1) \right] + 0$$

$$= -17.25/EI$$

$$f_{22} = \frac{1}{EI} \int m_2^2 \, dx = \frac{1}{EI}\left[-\frac{1}{2}(3)(3)(-2) + \frac{1}{2}(2)(2)\left(\frac{4}{3}\right)\right]$$

$$+ \frac{1}{2EI}\left[-3(3)(-3)\right] + \frac{1}{EI}\left[-\frac{1}{2}(3)(3)(-2)\right]$$

$$= +34.17/EI$$

$$f_{23} = f_{32} = \frac{1}{EI}\int \frac{m_2 m_3}{EI} = \frac{1}{EI}\left[-\frac{1}{2}(3)(3) + \frac{1}{2}(2)(2)\right](-1)$$

$$+ \frac{1}{2EI}\left[(-3)(3)(-1)\right] + \frac{1}{EI}\left[-\frac{1}{2}(3)(3)(-1)\right]$$

$$= +11.5/EI$$

$$f_{33} = \frac{1}{EI}\int m_3^2 \, dx = \frac{1}{EI}\left[-1(5)(-1)\right] + \frac{1}{2EI}\left[-1(3)(-1)\right] + \frac{1}{EI}\left[-1(3)(-1)\right]$$

$$= +9.5/EI$$

<u>Step 5</u>: Substitute in the equations of consistent deformation:

$$\begin{bmatrix} 49.5 & -14.25 & -17.25 \\ -14.25 & 34.17 & 11.5 \\ -17.25 & 11.5 & 9.5 \end{bmatrix} \begin{Bmatrix} R_1 \\ R_2 \\ R_3 \end{Bmatrix} = -\begin{Bmatrix} -136.25 \\ 8.25 \\ 45.75 \end{Bmatrix}$$

Solve the equations using the computer program <u>SOLVE</u>:

$\longrightarrow$ Solve ⟨enter⟩

Enter the number of equations:

$\longrightarrow$ 3 ⟨enter⟩

Enter the linear system one equation at a time:
Conclude each equation with its right-hand constant:
Enter coefficients of equation 1:

$\longrightarrow$ 49.5, −14.25, −17.25, 136.25 ⟨enter⟩

Enter coefficients of equation 2:

$\longrightarrow$ −14.25, 34.17, 11.5, −8.25 ⟨enter⟩

Enter coefficients of equation 3:

−17.25, 11.5, 9.5, −45.75 ⟨enter⟩

<u>Solution</u>:

$$R_1 = 2.358 \implies D_y = 2.36 \text{ kN} \uparrow$$
$$R_2 = 1.555 \implies D_x = 1.56 \text{ kN} \leftarrow$$
$$R_3 = -2.4169 \implies M_D = 2.42 \text{ kN·m}$$

Chapter 12

Method of Least Work
(Castigliano's Second Theorem)

12.1 Introduction:

This is a __force method__. (the unknowns are forces).
(it uses __redundants__).

* In this method, the strain __energy__ of the system must be determined and minimized.

* There is __no__ need to calculate deflections.

Menabréa 1858
Carlo Castigliano 1875
1879 (published)

12.2 Derivation of Castigliano's Theorems:

Case 1: $\delta_1 = f_{11} F_1$
          $\delta_{21} = f_{21} F_1$

Case 2:
          $\delta_1 = f_{12} F_2$
          $\delta_2 = f_{22} F_2$

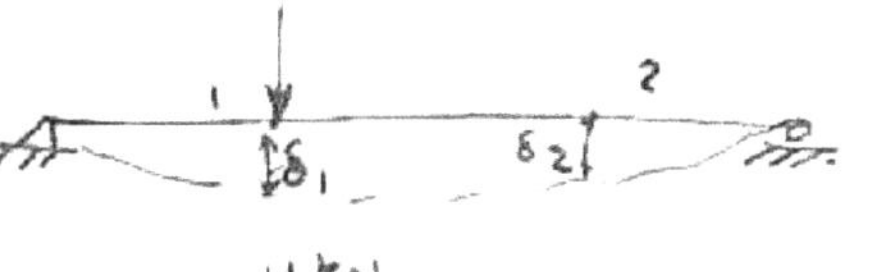

Calculate the work due to F_1 and F_2:

① Apply F_1 first: $dW_1 = \frac{1}{2} F_1 \delta_{11} = \frac{1}{2} F_1 (f_{11} F_1) = \frac{1}{2} f_{11} F_1^2$

Now apply F_2 while F_1 is still acting:

$$dW_2 = \frac{1}{2} F_2 \delta_{22} = \frac{1}{2} F_2 (f_{22} F_2) = \frac{1}{2} f_{22} F_2^2$$

Also $dW_3 = F_1 \delta_{12} = F_1 (f_{12} F_2)$ $\left(no\ \frac{1}{2}\right)$

$$= f_{12} F_1 F_2$$

$\therefore\ W = \frac{1}{2} f_{11} F_1^2 + \frac{1}{2} f_{22} F_2^2 + f_{12} F_1 F_2$

② Apply F_2 first. $dW_1 = \frac{1}{2} F_2 \delta_2 = \frac{1}{2} F_2 (f_{22} F_2) = \frac{1}{2} f_{22} F_2^2$

Now apply F_1 while F_2 is still acting.

$$dW_2 = \frac{1}{2} F_1 \delta_1 = \frac{1}{2} F_1 (f_{11} F_1) = \frac{1}{2} f_{11} F_1^2$$

$$dW_3 = F_2 \delta_{21} = F_2 (f_{21} F_1) = f_{21} F_1 F_2$$

$$W = \frac{1}{2} f_{22} F_2^2 + \frac{1}{2} f_{11} F_1^2 + f_{21} F_1 F_2$$

In a linear system, the work must be equal.

$$\tfrac{1}{2} f_{11} F_1^2 + \tfrac{1}{2} f_{22} F_2^2 + f_{12} F_1 F_2 = \tfrac{1}{2} f_{22} F_2^2 + \tfrac{1}{2} f_{11} F_1^2 + f_{21} F_2 F_1$$

$$f_{12} F_1 F_2 = f_{21} F_2 F_1$$

$$\boxed{f_{21} = f_{12}} \quad \text{Maxwell's law of Reciprocal deflection}$$

(Maxwell's Reciprocal theorem)

Note this theorem applies also to moments and rotations.
Thus the rotation at 1 due to a unit force at 2 is equal to the deflection at 2 due to a unit couple at 1

$$\delta_{12} = \delta_{21}$$

Apply the two forces F_1 and F_2 simultaneously —,

$$W = \tfrac{1}{2} F_1 \delta_1 + \tfrac{1}{2} F_2 \delta_2$$

where
$$\delta_1 = f_{11} F_1 + f_{12} F_2$$
$$\delta_2 = f_{21} F_1 + f_{22} F_2$$

$$W = \tfrac{1}{2} F_1 \left(f_{11} F_1 + f_{12} F_2 \right) + \tfrac{1}{2} F_2 \left(f_{21} F_1 + f_{22} F_2 \right)$$

$$= \tfrac{1}{2} f_{11} F_1^2 + \tfrac{1}{2} f_{12} F_1 F_2 + \tfrac{1}{2} f_{21} F_2 F_1 + \tfrac{1}{2} f_{22} F_2^2$$

But the strain energy $U = $ work (external) $= W$

$$= \tfrac{1}{2} f_{11} F_1^2 + \tfrac{1}{2} f_{12} F_1 F_2 + \tfrac{1}{2} f_{21} F_2 F_1 + \tfrac{1}{2} f_{22} F_2^2$$

but $f_{12} = f_{21}$

$$U = \tfrac{1}{2} f_{11} F_1^2 + f_{12} F_1 F_2 + \tfrac{1}{2} f_{22} F_2^2$$

$$\frac{\partial U}{\partial F_1} = f_{11} F_1 + f_{12} F_2 = \delta_1$$

$$\frac{\partial U}{\partial F_2} = f_{12} F_1 + f_{22} F_2 = \delta_2$$

$$\boxed{\frac{\partial U}{\partial F_1} = \delta_1} \implies \text{Castigliano's First Theorem}$$

<u>Theorem 1</u> : "The partial derivative of the strain energy in a structure with respect to one of the external forces acting on the structure is equal to the displacement (deflection) at that force in the direction of the force"

$$\delta_1 = \frac{\partial U}{\partial P_1}$$

$$\delta_2 = \frac{\partial U}{\partial P_2}$$

However, $\frac{\partial U}{\partial R_1} = \delta_{R_1} = 0$ (deflection is zero at the support)

$$\therefore \boxed{\frac{\partial U}{\partial R_1} = 0} \implies \text{Castigliano's Second Theorem}$$
$$\underline{\text{(Method of Least Work)}}.$$

the strain energy is <u>minimized</u> (least work).

<u>Theorem 2</u> : "Redundants must have a value that will make the strain energy in a structure a minimum".

12.3 <u>Application of the Method of Least Work</u>:

(1) Choose the redundants.
(2) Write the strain energy function U in terms of the redundants.
(3) Minimize it for <u>each</u> redundant.
(4) Obtain & Solve the system of simultaneous equations.

(1) For a beam or a frame:
$$U = \int \frac{1}{2} \frac{M^2 dx}{EI} \implies \frac{\partial U}{\partial R} = \int \frac{1}{2EI} \cdot 2M \frac{\partial M}{\partial R} dx$$
$$= \int \frac{M}{EI} \frac{\partial M}{\partial R} dx$$

(2) For a truss (or cable):
$$U = \sum \frac{P^2 L}{2EA} \implies \frac{\partial U}{\partial R} = \sum \frac{1}{2EA} 2P \frac{\partial P}{\partial R} L = \sum \frac{PL}{EA} \frac{\partial P}{\partial R}$$

Example 12.1

Determine the reactions for the propped cantilever shown in the figure?

Solution: We have one redundant

Choose R_A as the redundant

$$\frac{\partial U}{\partial R_A} = \int \frac{M}{EI}\frac{\partial M}{\partial R_A}\,dx = 0$$

$0 \leq x \leq L/2$:

$$M = R_A x \quad , \quad \frac{\partial M}{\partial R_A} = x$$

$L/2 \leq x \leq L$:

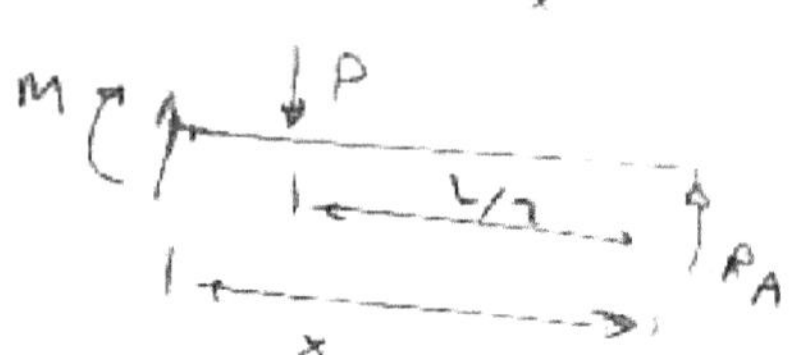

$$M = R_A x - P\left(x - \frac{L}{2}\right)$$

$$= R_A x - P x + P\frac{L}{2}$$

$$= (R_A - P)x + PL/2$$

$$\frac{\partial M}{\partial R_A} = x$$

$$\frac{\partial U}{\partial R_A} = \int \frac{M}{EI}\frac{\partial M}{\partial R_A}\,dx = \frac{1}{EI}\int_0^{L/2}(R_A x)(x)\,dx + \frac{1}{EI}\int_{L/2}^{L}\left[(R_A - P)x + \frac{PL}{2}\right](x)\,dx = 0$$

$$\frac{1}{EI}\left[R_A\frac{x^3}{3}\right]_0^{L/2} + \frac{1}{EI}\left[(R_A - P)\frac{x^3}{3} + \frac{PL}{2}\frac{x^2}{2}\right]_{L/2}^{L} = 0$$

$$\frac{1}{3}R_A\left(\frac{L}{2}\right)^3 + \frac{1}{3}(R_A - P)(L)^3 + \frac{PL}{4}(L)^2 - \frac{1}{3}(R_A - P)\left(\frac{L}{2}\right)^3 - \frac{PL}{4}\left(\frac{L}{2}\right)^2 = 0$$

$$\frac{1}{24}R_A L^3 + \frac{1}{3}R_A L^3 - \frac{1}{3}PL^3 + \frac{1}{4}PL^3 - \frac{1}{24}R_A L^3 + \frac{1}{24}PL^3 - \frac{1}{16}PL^3 = 0$$

$$\frac{1}{3} R_A L^3 + \left(-\frac{1}{3} + \frac{1}{4} + \frac{1}{24} - \frac{1}{16}\right) P L^3 = 0$$

$$\frac{1}{3} R_A L^3 + \left(\frac{-16 + 12 + 2 - 3}{48}\right) P L^3 = 0$$

$$\frac{1}{3} R_A + \left(-\frac{5}{48}\right) P = 0 \quad \Longrightarrow \quad \boxed{R_A = \frac{5P}{16}}$$

Equations of Equilibrium:

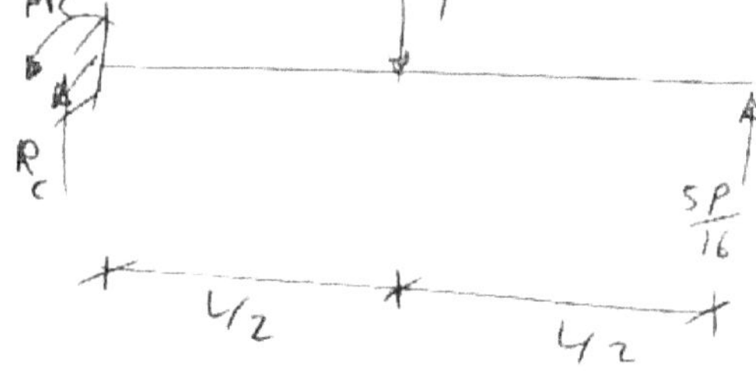

$+\uparrow \Sigma F_y = 0$:

$$R_c + \frac{5}{16} P - P = 0$$

$$R_c = \frac{11}{16} P \uparrow$$

$+\circlearrowleft \Sigma M_c = 0$:

$$M_c + \frac{5}{16} P(L) - P\left(\frac{L}{2}\right) = 0$$

$$M_c = \frac{1}{2} PL - \frac{5}{16} PL = \frac{3}{16} PL \quad \circlearrowleft$$

Example 12.2:

Determine the reactions for the continuous beam shown in the figure.

Solution: $R_{A_x} = 0$:

We have __one__ redundant.

Choose R_A as the redundant.

$$\boxed{\frac{\partial U}{\partial R_A} = \int \frac{M}{EI} \frac{\partial M}{\partial R_A} dx = 0}$$

$\underline{0 \le x \le 10}$

$$M = R_A x$$

$$\frac{\partial M}{\partial R_A} = x$$

$\underline{0 \le x \le 20}:$

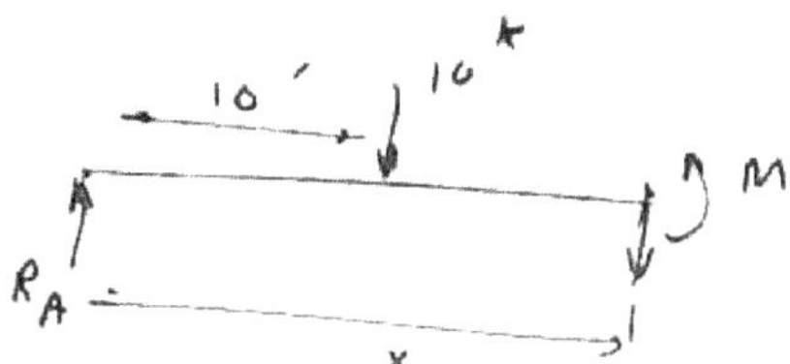

$$M = R_A x - 10(x - 10)$$

$$\frac{\partial M}{\partial R_A} = x$$

$\underline{0 \leq x \leq 10}$:

$$M = R_D x - 2x\left(\frac{x}{2}\right) = R_D x - x^2$$

We have to write R_D in terms of R_A:

use the equations of equilibrium.

$+\circlearrowleft \; \Sigma M_c = 0$:

$$R_D(10) - 2(10)(5) + 10(10) - R_A(20) = 0$$

$$10 R_D - 100 + 100 - 20 R_A = 0$$

$$R_D = 2 R_A$$

$$\therefore \quad M = 2 R_A x - x^2$$

$$\frac{\partial M}{\partial R_A} = 2x$$

$$\frac{\partial U}{\partial R_A} = \int \frac{M}{EI}\frac{\partial M}{\partial R_A}dx = \frac{1}{EI}\int_0^{10}(R_A x)x\,dx + \frac{1}{EI}\int_{10}^{20}\left(R_A x - 10(x-10)\right)x\,dx + \frac{1}{EI}\int_0^{10}(2R_A x - x^2)(2x\,dx) = 0$$

$$R_A\int_0^{10}x^2 dx + R_A\int_{10}^{20}x^2 dx - 10\int_{10}^{20}(x-10)x\,dx + 4R_A\int_0^{10}x^2 dx - 2\int_0^{10}x^3 dx = 0$$

$$R_A\left[\frac{x^3}{3}\right]_0^{10} + R_A\left[\frac{x^3}{3}\right]_{10}^{20} - 10\left[\frac{x^3}{3} - \frac{10}{2}x^2\right]_{10}^{20} + 4R_A\left[\frac{x^3}{3}\right]_0^{10} - 2\left[\frac{x^4}{4}\right]_0^{10} = 0$$

$$R_A\left(\frac{1000}{3}\right) + R_A\left(\frac{8000}{3} - \frac{1000}{3}\right) \quad 10\left[\frac{8000}{3} - 5(400) - \frac{1000}{3} + 5(100)\right] \quad \boxed{123}$$

$$+ 4R_A\left(\frac{1000}{3}\right) - \frac{1}{2}(10,000) = 0$$

$$\frac{8000}{3}R_A + \frac{4000}{3}R_A = 10\left[\frac{2000}{3} - 300(5)\right] + 5,000$$

$$4,000\,R_A = 13,333.33 \quad \Longrightarrow \quad R_A = 3.33 \text{ k} \uparrow$$

Equations of equilibrium.

$$R_D = 2R_A = 2(3.33) = 6.66 \text{ k} \uparrow$$

$+\uparrow \ \Sigma F_y = 0$

$$3.33 + R_c + 6.66 - 10 - 2(10) = 0$$

$$R_c = 20 \text{ k} \uparrow$$

Example 12.3:

Determine the reactions for the frame shown in the figure. EI = constant.

Solution:

We have __two__ redundants.

Choose V_A and H_A as the redundants.

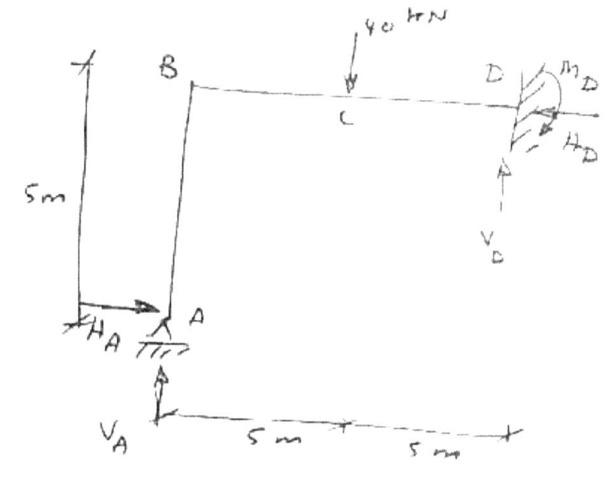

$$\boxed{\begin{array}{l} \dfrac{\partial U}{\partial H_A} = \displaystyle\int \dfrac{M}{EI}\dfrac{\partial m}{\partial H_A}dx = 0 \quad\text{———— (1)} \\[3mm] \dfrac{\partial U}{\partial V_A} = \displaystyle\int \dfrac{M}{EI}\dfrac{\partial m}{\partial V_A}dx = 0 \quad\text{———— (2)} \end{array}}$$

$0 \le x \le 5$:

$$M = -H_A x$$

$$\frac{\partial m}{\partial H_A} = -x$$

$$\frac{\partial m}{\partial V_A} = 0$$

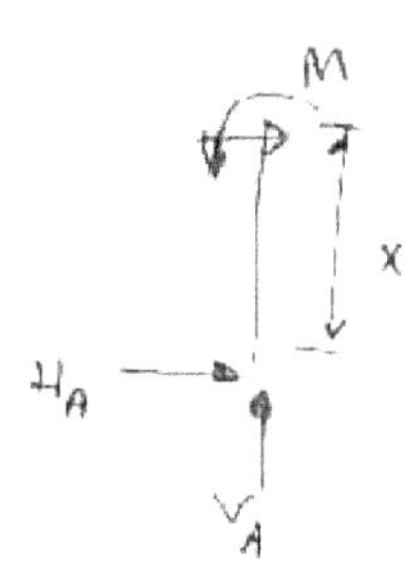

$\underline{0 \leq x \leq 5}$

$M = -5H_A + V_A x$

$\dfrac{\partial M}{\partial H_A} = -5$

$\dfrac{\partial M}{\partial V_A} = x$

$\underline{5 \leq x \leq 10}$

$M = -5H_A + V_A x - 40(x-5)$

$\dfrac{\partial M}{\partial H_A} = -5$

$\dfrac{\partial M}{\partial V_A} = x$

$\dfrac{\partial U}{\partial H_A} = \dfrac{1}{EJ}\displaystyle\int_0^5 (-H_A x)(-x)\,dx + \dfrac{1}{EJ}\int_0^5 (-5H_A + V_A x)(-5)\,dx + \dfrac{1}{EI}\int_5^{10}(-5H_A + xV_A - 40(x-5))(-5)\,dx = 0 \quad (1)$

$\dfrac{\partial U}{\partial V_A} = 0 + \dfrac{1}{EJ}\displaystyle\int_0^5 (-5H_A + V_A x)(x)\,dx + \dfrac{1}{EI}\int_5^{10}(-5H_A + V_A x - 40(x-5))(x)\,dx = 0 \quad (2)$

$H_A \displaystyle\int_0^5 x^2\,dx + 5\int_0^5 (5H_A - xV_A)\,dx + 5\int_5^{10}(5H_A - xV_A + 40(x-5))\,dx = 0 \qquad (1)$

$\displaystyle\int_0^5 x(-5H_A + xV_A)\,dx + \int_5^{10} x(-5H_A + V_A x - 40(x-5))\,dx = 0 \qquad (2)$

$H_A\left[\dfrac{x^3}{3}\right]_0^5 + 5\left[5H_A x - V_A \dfrac{x^2}{2}\right]_0^5 + 5\left[5H_A x - V_A \dfrac{x^2}{2} + 40\left(\dfrac{x^2}{2} - 5x\right)\right]_5^{10} = 0 \qquad (1)$

$\left[-5H_A \dfrac{x^2}{2} + V_A \dfrac{x^3}{3}\right]_0^5 + \left[-5H_A \dfrac{x^2}{2} + V_A \dfrac{x^3}{3} - 40\left(\dfrac{x^3}{3} - 5\dfrac{x^2}{2}\right)\right]_5^{10} = 0 \qquad (2)$

$$H_A\left[\frac{(5)^3}{3}\right] + 5\left[5H_A(5) - V_A\frac{(5)^2}{2}\right] + 5\left[5H_A(10) - V_A\frac{(10)^2}{2} + 40\left(\frac{(10)^2}{2} - 5(10)\right)\right]$$

$$-5\left[5H_A(5) - V_A\frac{(5)^2}{2} + 40\left(\frac{(5)^2}{2} - 5(5)\right)\right] = 0 \qquad\text{—— (1)}$$

$$-5H_A\frac{(5)^2}{2} + V_A\frac{(5)^3}{3} + \left[-5H_A\frac{(10)^2}{2} + V_A\frac{(10)^3}{3} - 40\left(\frac{(10)^3}{3} - 5\frac{(10)^2}{2}\right) + 40\left(\frac{(5)^3}{3} - 5\frac{(5)^2}{2}\right)\right]$$

$$+ 5H_A\frac{(5)^2}{2} - V_A\frac{(5)^3}{3} = 0 \qquad\text{—— (2)}$$

$$\left\{\begin{array}{l} 291.67H_A - 250V_A + 2500 = 0 \qquad\text{—— (1)} \\[2mm] -250H_A + 333.33V_A - 4166.67 = 0 \qquad\text{—— (2)} \end{array}\right\}$$

Solve eqs. (1) & (2) simultaneously for H_A and V_A:

Multiply (1) by 1.333:

$$388.89H_A - \cancel{333.33}V_A + 3333.33 = 0 \qquad\text{—— (1)}$$

$$-250H_A + \cancel{333.33}V_A - 4166.67 = 0 \qquad\text{—— (2)}$$

$$\Rightarrow \quad H_A = 6.0 \text{ kN} \longrightarrow$$

$$\therefore \quad V_A = 17.0 \text{ kN} \uparrow$$

Equations of Equilibrium:

$\xrightarrow{+} \Sigma F_x = 0:$
$$-H_D + 6 = 0 \Rightarrow H_D = 6 \text{ kN} \leftarrow$$

$+\uparrow \Sigma F_y = 0: \quad 17 - 40 + V_D = 0$
$$V_D = 23 \text{ kN} \uparrow$$

$+\circlearrowright \Sigma M_D = 0: \quad -M_D + 40(5) + 6(5) - 17(10) = 0$
$$M_D = 60 \text{ kN·m} \circlearrowright$$

Example 12.4

Determine the reactions in the wires that support the beam shown in the figure. Each wire has a cross-sectional area of $0.1 \, in^2$, $I = 288 \, in^4$ for the beam, and $E = 30 \times 10^6 \, psi$ for both the wires and the beam.

Solution:

We have one redundant.
Choose T_A as the redundant.

From symmetry,
$$T_c = T_A .$$

$+\circlearrowleft \Sigma M_c = 0:$

$$10(10) - T_B(10) - T_A(20) = 0$$

$$20 T_A = 100 - 10 T_B$$

$$T_B = 10 - 2 T_A$$

$$\frac{\partial U}{\partial T_A} = \int \frac{M}{EI} \frac{\partial m}{\partial T_A} dx \; + \; \Sigma \frac{TL}{EA} \frac{\partial T}{\partial T_A} \qquad = 0$$

$0 \le x \le 10 :$

$$M = T_A x$$

$$\frac{\partial m}{\partial T_A} = x$$

$$\therefore \frac{\partial U}{\partial T_A} = 2 \int \left(\frac{T_A x}{EI}\right)(x) dx \; + \; \frac{2 T_A L}{EA}\left(\frac{\partial T_A}{\partial T_A}\right) + \; \frac{T_B L}{EA}\left(\frac{\partial T_B}{\partial T_A}\right) \qquad = 0$$

$$\frac{2 T_A}{EI} \int_0^{10} x^2 dx \; + \; \frac{2 T_A L}{EA}(1) \; + \; \frac{(10 - 2 T_A) L}{EA}(-2) = 0$$

$$\frac{2 T_A}{EI}\left(\frac{10^3}{3}\right) + \frac{2 T_A (5)}{EA} \; + \; \frac{(10 - 2 T_A)(5)(-2)}{EA} = 0$$

$$\frac{2 \times 10^3 \, T_A}{3 \, I} \; + \; \frac{10 T_A}{A} \; - \; \frac{10(10 - 2 T_A)}{A} = 0$$

take $A = 0.1 \, in^2 = \dfrac{0.1}{(12)^2} = \dfrac{0.1}{144}$

$$I = 288 \, in^4 = \frac{288}{(12)^4}$$

$$\frac{2 \times 10^3}{3} \cdot \frac{T_A (12)^4}{288} + \frac{10 \, T_A (12)^2}{0.1} - \frac{20 (5 - T_A)(12)^2}{0.1} = 0$$

$$\frac{2 \times 10^3}{3 \times 288}(12)^2 T_A + \frac{10}{0.1} T_A - \frac{20(5 - T_A)}{0.1} = 0$$

$$\left(\frac{2 \times 10^3 \times 144}{3 \times 288} + \frac{10}{0.1} + \frac{20}{0.1} \right) T_A = \frac{100}{0.1}$$

$$\implies T_A = 1.58 \, k$$

$$T_c = T_A = 1.58 \, k$$

$$T_B = 10 - 2 T_A = 10 - 2(1.58) = 6.84 \, k$$

Example 12.5 :

Determine the member forces for the truss shown in the figure. E = constant

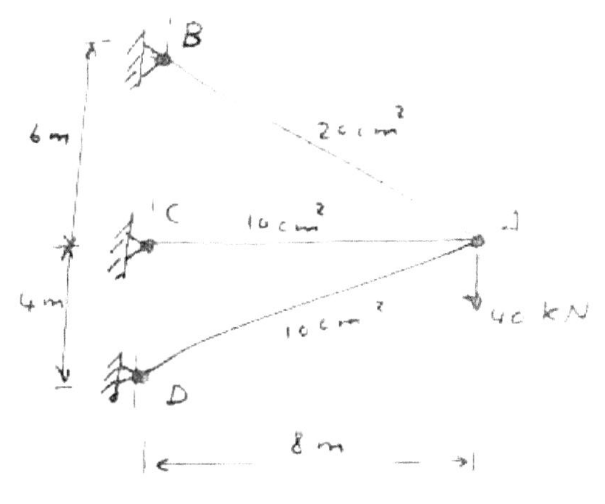

Solution:

$$J = 4 \implies 2j = 8$$

$$r = 6 \quad , \quad m = 3$$

$$r + m = 9$$

$$\implies 2j > r + m, \quad 9 - 8 = 1$$

∴ The truss is indeterminate to the first degree.

We have one redundant.

Choose F_{AB} as the redundant.

$$\frac{\partial U}{\partial F_{AB}} = \sum \frac{F L}{E A} \left(\frac{\partial F}{\partial F_{AB}} \right)$$

Joint A :

$+\uparrow \Sigma F_y = 0 :$

$$F_{AB}\left(\frac{6}{10}\right) - 40 - F_{AD}\left(\frac{4}{8.94}\right) = 0$$

$$\boxed{F_{AD} = 1.341 \, F_{AB} - 89.4} \qquad \frac{\partial F_{AD}}{\partial F_{AB}} = 1.341$$

$$\Sigma F_x = 0 \qquad -F_{AC} - F_{AB}\left(\frac{8}{10}\right) - F_{AD}\left(\frac{8}{8.94}\right) = 0$$

$$F_{AC} = -0.8 F_{AB} - \frac{8}{8.94}\left(1.341 F_{AB} - 89.4\right)$$

$$\boxed{F_{AC} = 80 - 2F_{AB}} \qquad \frac{\partial F_{AC}}{\partial F_{AB}} = -2$$

Member	$A\ (cm^2)$	$L\ (m)$	F	$\partial F/\partial F_{AB}$	$\frac{FL}{A}\frac{\partial F}{\partial F_{AB}}$
AB	20	10	F_{AB}	1	$0.5 F_{AB}$
AC	10	8	$80 - 2F_{AB}$	-2	$-1.6(80 - 2F_{AB})$
AD	10	8.94	$1.341 F_{AB} - 89.4$	1.341	$1.1988(1.341 F_{AB} - 89.4)$

$$\frac{\partial U}{\partial F_{AB}} = \Sigma \frac{FL}{EA}\frac{\partial F}{\partial F_{AB}} = \frac{1}{E}\Sigma \frac{FL}{A}\frac{\partial F}{\partial F_{AB}} = 0$$

$$0.5 F_{AB} - 1.6(80 - 2F_{AB}) + 1.1988(1.341 F_{AB} - 89.4) = 0$$

$$F_{AB} = 44.3 \ kN \ (T).$$

$$\therefore F_{AC} = 80 - 2F_{AB} = 80 - 2(44.3) = -8.6 \ kN \ (C).$$

$$F_{AD} = 1.341 F_{AB} - 89.4 = 1.341(44.3) - 89.4 = -29.99 \ kN$$
$$\simeq -30 \ kN \ (C).$$

Example 12.6:

Draw the bending moment diagram for the ring (closed ring) shown in the figure.

Solution:

*A closed ring is an indeterminate structure.

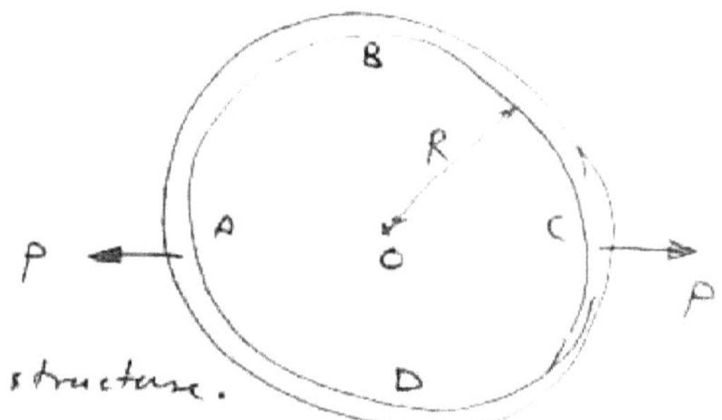

Since the ring is symmetric about a vertical line through B and D, there is no shear at B and D.

The equations of equilibrium were already used, and one still have one unknown.

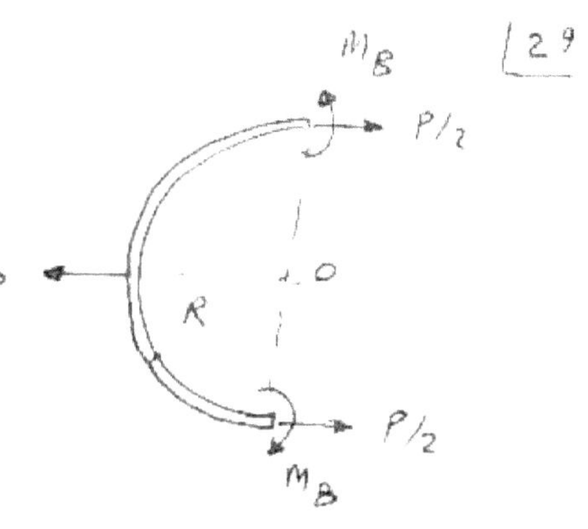

We have __one__ redundant, M_B.

$$\frac{\partial U}{\partial M_B} = \int \frac{M}{EI}\frac{\partial M}{\partial M_B}\,dx = 0$$

we can use $dx \approx ds$ for __thin__ rings

$+\circlearrowleft\ \Sigma M_x = 0:$

$$-M + M_B - \frac{P}{2}(R - R\cos\theta) = 0$$

$$M = M_B - \frac{PR}{2}(1 - \cos\theta)$$

$$\frac{\partial M}{\partial M_B} = 1$$

$$ds = R\,d\theta$$

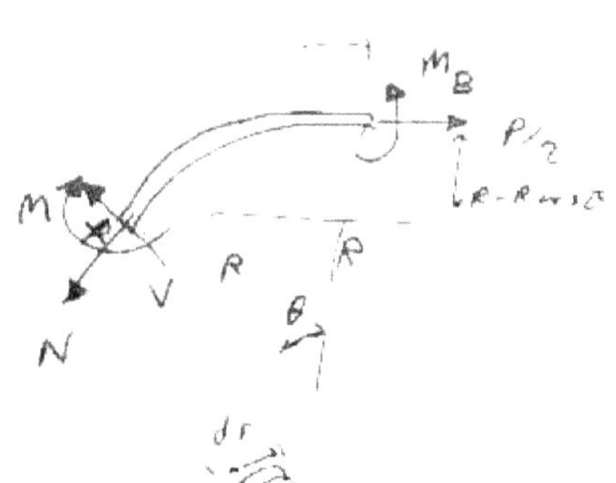

$$\frac{\partial U}{\partial M_B} = \int \frac{M}{EI}\frac{\partial M}{\partial M_B}\,ds = 0$$

$$= \frac{4}{EI}\int_0^{\pi/2}\left[\left(M_B - \frac{1}{2}PR(1-\cos\theta)\right)\right](1)\cdot R\,d\theta = 0$$

$$= \frac{4R}{EI}\int_0^{\pi/2}\left[M_B - \frac{1}{2}PR(1-\cos\theta)\right]d\theta = 0$$

$$= \frac{4R}{EI}\left[M_B\theta - \frac{1}{2}PR(\theta - \sin\theta)\right]_0^{\pi/2} = 0$$

$$= \frac{4R}{EI}\left[M_B\frac{\pi}{2} - \frac{1}{2}PR\left(\frac{\pi}{2}-1\right) - 0 + \frac{1}{2}PR(0-0)\right] = 0$$

$$\frac{4R}{EI}\left[\frac{\pi}{2}M_B - \frac{1}{2}PR\left(\frac{\pi}{2}-1\right) + 0\right] = 0$$

$$M_B = \frac{2}{\pi}\cdot\frac{1}{2}PR\left(\frac{\pi}{2}-1\right) \implies M_B = \frac{PR}{\pi}\left(\frac{\pi}{2}-1\right) = \underline{0.182\ PR}$$

$$M_{\frac{\pi}{2}} = M_A = 0.182\,PR - \frac{PR}{2}\left(1-\cos\frac{\pi}{2}\right) = \underline{-0.318\ PR}$$

B.M.D.

The Three-Moment Equation

* This is a force method (the unknowns are moments).
* Clapeyron (French, 1857), applicable only to beams.
* Choose M_i, M_j, M_k as the redundants

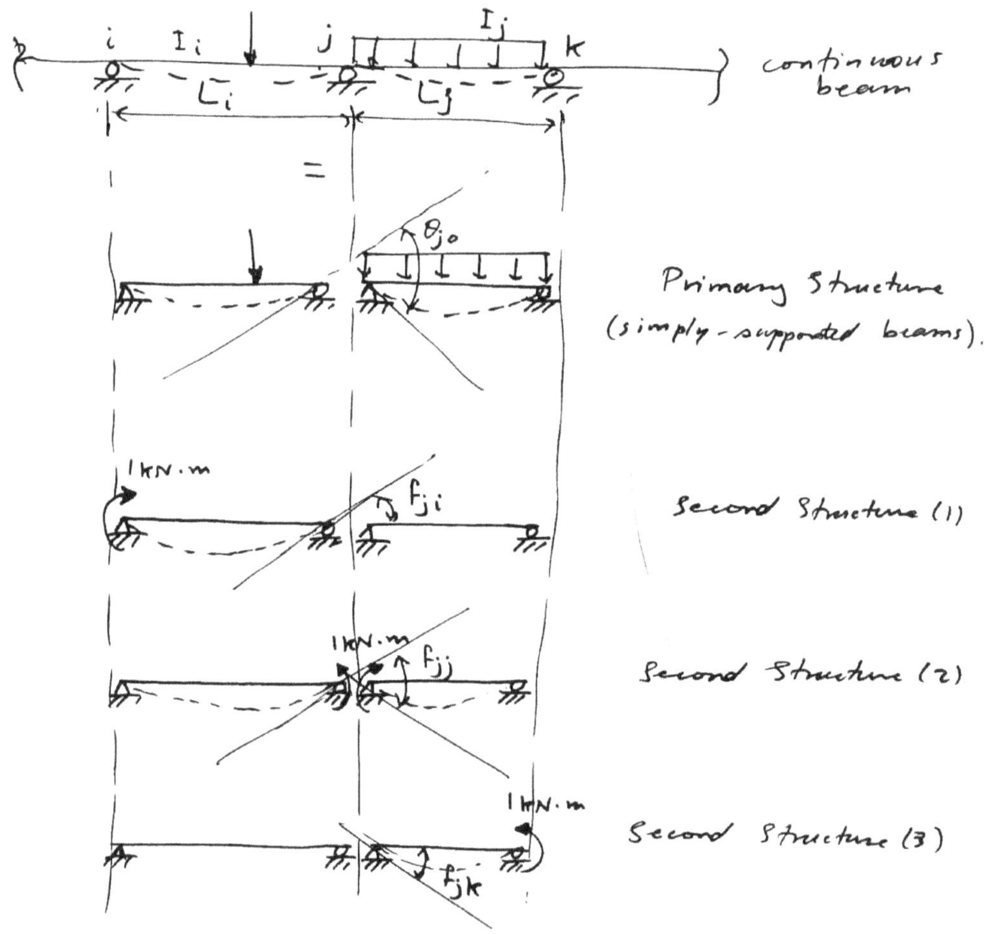

Substitute in the equation of consistent deformations:

$$f_{ji} M_i + f_{jj} M_j + f_{jk} M_k + \theta_{jo} = \theta_j \equiv 0$$

Calculate all the required deflections f_{ji}, f_{jj}, f_{jk}, θ_{jo}
in the structures using any method for calculating deflections.

We have:

$$f_{ji} = \frac{L_i}{6 E I_i}$$

$$f_{jj} = \frac{L_i}{3 E I_i} + \frac{L_j}{3 E I_j}$$

$$f_{jk} = \frac{L_j}{6 E I_j}$$

$$\theta_{jo} = \ ? \quad \text{(obtained from the primary structure)} \\ \text{(depends on the loads).}$$

$$\frac{L_i}{6EI_i}\,M_i + \left(\frac{L_i}{3EI_i} + \frac{L_j}{3EI_j}\right)M_j + \frac{L_j}{6EI_j}\,M_k = -\,\theta_{jo}$$

$$\therefore \quad \left(\frac{L_i}{I_i}\right)M_i + \left(2\frac{L_i}{I_i} + 2\frac{L_j}{I_j}\right)M_j + \left(\frac{L_j}{I_j}\right)M_k = -\,6E\,\theta_{jo}$$

*   The right-hand side depends on the load on the beam.

*   We will derive θ_{jo} from the primary structure for different types of loads.

Case 1 :

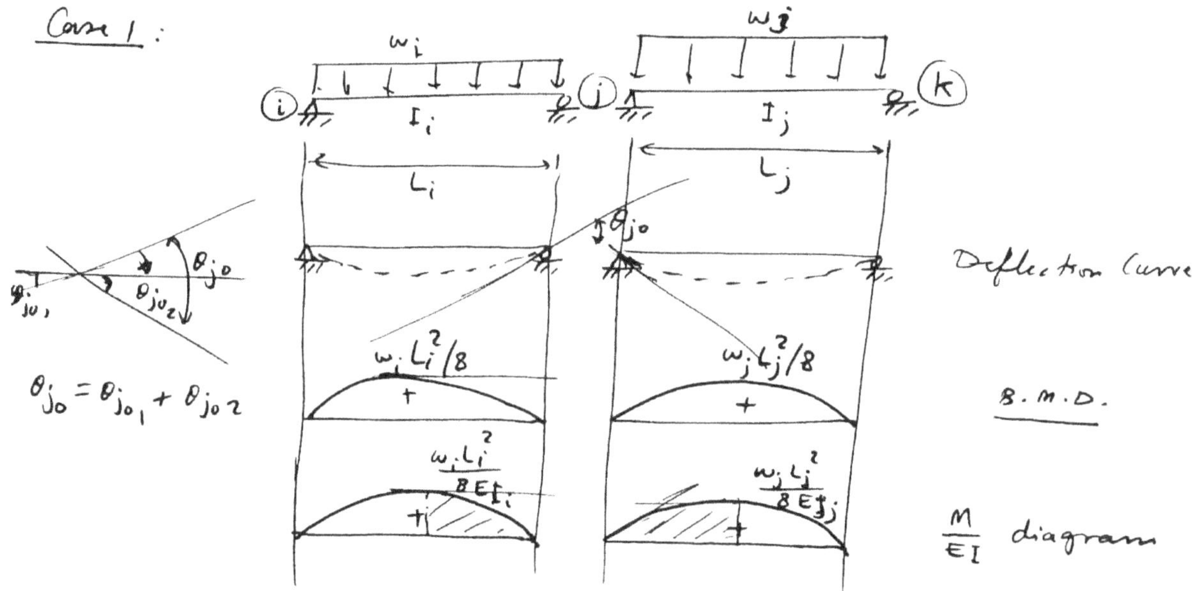

Using the moment-area theorems.

$$\theta_{jo_1} = \text{shaded area} = \frac{2}{3}\left(\tfrac{1}{2}L_i\right)\left(\frac{w_i L_i^2}{8EI_i}\right) = \frac{w_i L_i^3}{24EI_i}\ .$$

$$\theta_{jo_2} = \text{shaded area} = \frac{2}{3}\left(\tfrac{1}{2}L_j\right)\left(\frac{w_j L_j^2}{8EI_j}\right) = \frac{w_j L_j^3}{24EI_j}\ .$$

$$\therefore \ \theta_{jo} = \theta_{jo_1} + \theta_{jo_2} = \frac{w_i L_i^3}{24EI_i} + \frac{w_j L_j^3}{24EI_j}$$

$\therefore$ the right-hand side is.

$$-\,6E\,\theta_{jo} = \boxed{-\frac{w_i L_i^3}{4 I_i} - \frac{w_j L_j^3}{4 I_j}}$$

$\therefore$ For each uniformly-distributed load on each span, add the term $\boxed{\dfrac{-wL^3}{4I}}$ to the right-hand side of the equation.

Case 2 : For concentrated loads:

Determine θ_{jo} using the conjugate beam method:

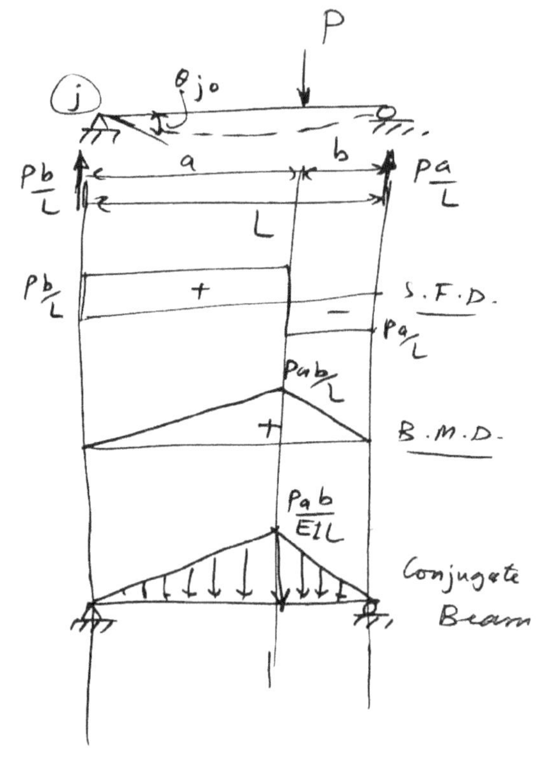

θ_{jo} = shear at left support

$\quad$ = left reaction

$\quad = \dfrac{1}{2}\left(\dfrac{Pab}{EIL}\right)(b)\left(\dfrac{2}{3}\dfrac{b}{L}\right)$

$\qquad + \dfrac{1}{2}\left(\dfrac{Pab}{EIL}\right)(a)\left(b+\dfrac{1}{3}a\right)\dfrac{1}{L}$

$= \dfrac{1}{2}\left(\dfrac{Pab}{EIL}\right)\left[\dfrac{2}{3}\dfrac{b^2}{L} + \dfrac{ab}{L} + \dfrac{a^2}{3L}\right]$

$= \dfrac{Pab}{2EIL}\left[\dfrac{2b^2 + 3ab + a^2}{3L}\right]$

$= \dfrac{Pab}{6EIL^2}\left(a^2 + 2b^2 + 3ab\right)$

$\therefore$ the right-hand side becomes:

$$-6EI\theta_{jo} = \boxed{-\dfrac{Pab}{IL^2}\left(a^2 + 2b^2 + 3ab\right)}$$

Note :
$\boxed{\begin{array}{l} a : \text{ \underline{near distance}} \\ b : \text{ \underline{far distance}} \end{array}}$

Case 3 : Concentrated Load P at midspan.

$\qquad a = b = \dfrac{L}{2}$

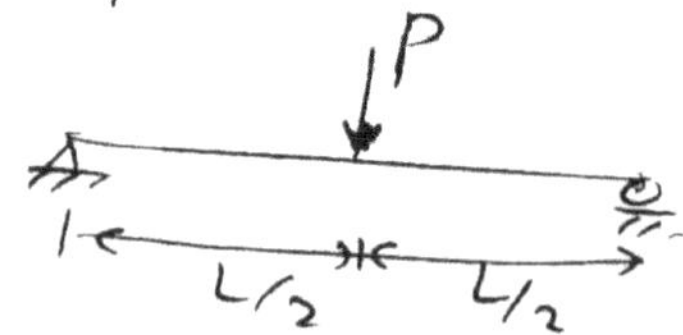

$\therefore$

$-6EI\theta_{jo} = -\dfrac{P\left(\frac{L}{2}\right)\left(\frac{L}{2}\right)}{IL^2}\left(\dfrac{L^2}{4} + 2\dfrac{L^2}{4} + 3\dfrac{L^2}{4}\right)$

$\qquad = -\dfrac{P}{4I}\left(\dfrac{6L^2}{4}\right)$

$-6EI\theta_{jo} = \boxed{-\dfrac{3PL^2}{8I}}$.

Procedure:

* For a continuous beam with m spans, we have $(m-1)$ interior supports $\Rightarrow$ $(m-1)$ redundants.

* Thus we have $(m-1)$ redundants.

* Thus we need $(m-1)$ equations.

* Write <u>one</u> equation for every <u>two</u> adjacent spans.

* For fixed ends, assume an imaginary end span with $L=0$ (or with $I=\infty$).

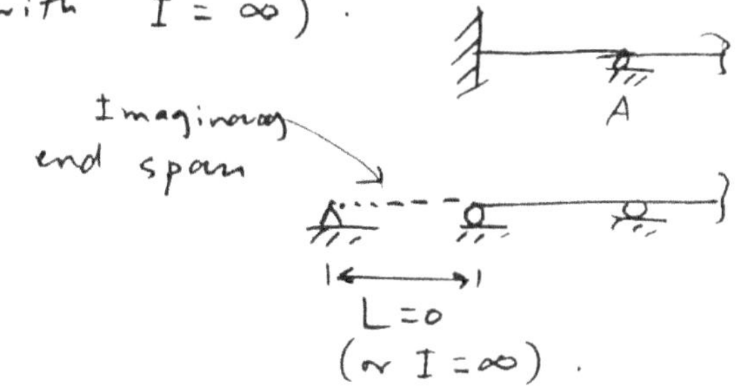

Example 1:

Draw the bending moment diagram for the structure shown in the figure.

Solution:

We have one redundant.

We have only <u>two</u> spans.

∴ We have $2-1 = 1$ equation.

Take $ABC \rightarrow i\,j\,k$:

Note: $\boxed{M_i = 0, \quad M_k = 0}$ (simple supports)

$$\left(\frac{L_i}{I_i}\right)M_i + 2\left(\frac{L_i}{I_i} + \frac{L_j}{I_j}\right)M_j + \left(\frac{L_j}{I_j}\right)M_k = -6E\theta_{jo}$$

In this case, $\quad 6E\theta_{jo} = \dfrac{w_j\,L_j^3}{4 I_j}$

∴
$$\left(\frac{L_i}{I_i}\right)M_i + 2\left(\frac{L_i}{I_i} + \frac{L_j}{I_j}\right)M_j + \left(\frac{L_j}{I_j}\right)M_k = -\frac{w_j\,L_j^3}{4 I_j}$$

$$\left(\frac{10}{I}\right)(0) + 2\left(\frac{10}{I} + \frac{20}{2I}\right)M_j + \left(\frac{20}{2I}\right)(0) = -\frac{5(20)^3}{4(2I)}$$

$$2\left(\frac{20}{I}\right)M_j = -\frac{5000}{I} \implies M_j = \underline{-125\ \text{k-ft}}$$

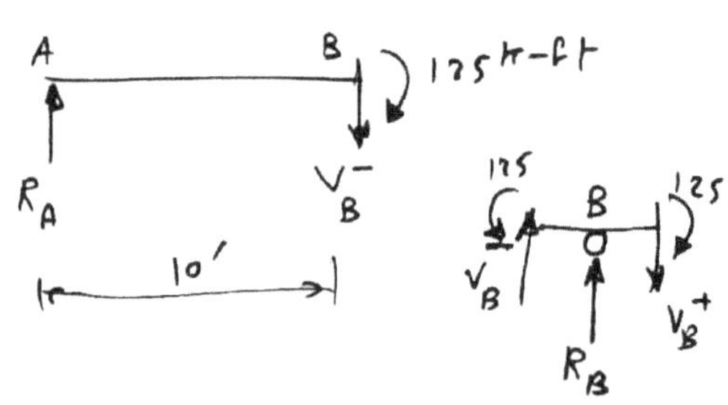

For AB :

$+\circlearrowleft \Sigma M_A = 0: \quad -V_B^-(10) - 125 = 0 \implies V_B^- = -12.5 \text{ k}.$

$+\uparrow \Sigma F_y = 0: \quad R_A - (-12.5) = 0 \implies R_A = -12.5 \text{ k}.$

$\therefore R_A = 12.5 \text{ k} \downarrow .$

For BC :

$+\circlearrowleft \Sigma M_C = 0: \quad 125 + 5(20)(10) - V_B^+(20) = 0$

$\qquad V_B^+ = 56.25 \text{ k}.$

$+\uparrow \Sigma F_y = 0: \quad 56.25 + R_C - 5(20) = 0 \implies R_C = 43.75 \text{ k} \uparrow .$

Support B :

$+\uparrow \Sigma F_y = 0: \quad V_B^- + R_B - V_B^+ = 0$

$\qquad -12.5 + R_B - 56.25 = 0 \implies R_B = 68.75 \text{ k} \uparrow .$

Locate point O of zero shear:

$$\frac{x}{20-x} = \frac{56.25}{43.75}$$

$$43.75x = 20(56.25) - 56.25x$$

$$\therefore x = 11.25'.$$

$$M_O = -125 + \frac{1}{2}(56.25)(11.25)$$

$$= +191.4 \text{ k-ft}.$$

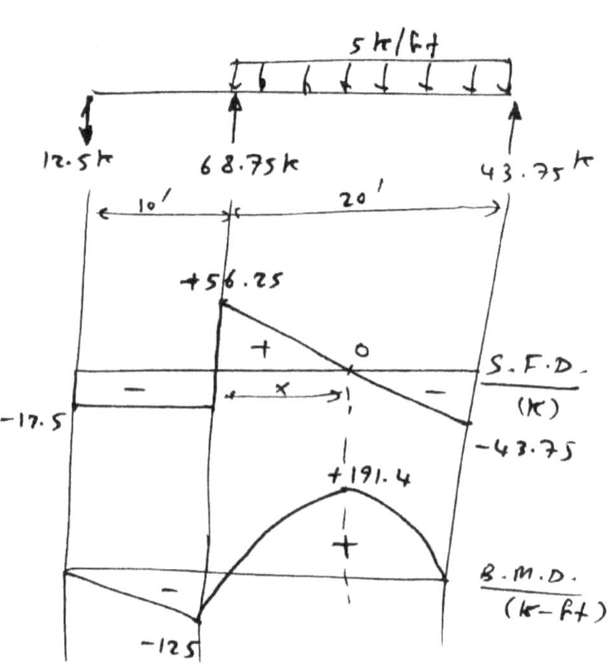

Example 2:

Determine the support moments for the structure shown in the figure.

Solution:

We have **two** redundants.

We have $2+1$ spans.

"imaginary span at the fixed end".

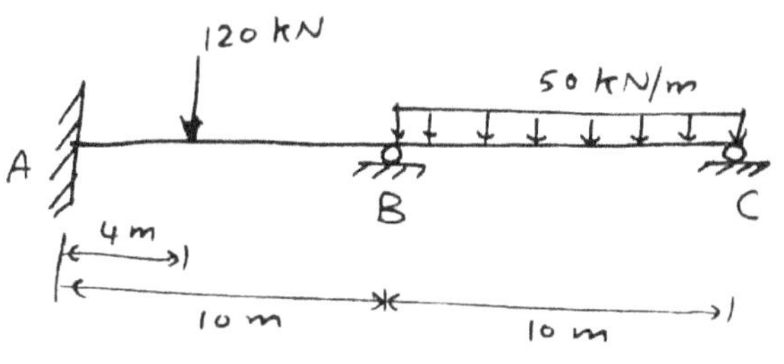

$\therefore m = 3 \implies m - 1 = 2 \implies$ We have **two** equations.

the redundants are the moments at the supports, M_A, M_B.

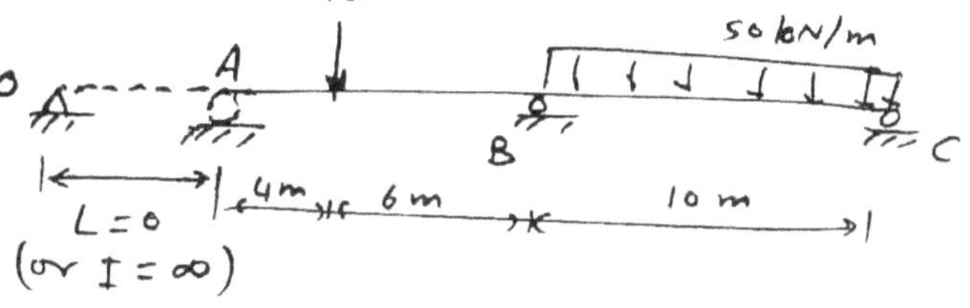

Equation (1) $\longrightarrow$ (i)   (j)

Equation (2) $\longrightarrow$ (i)   (j)

Equation (1): $OAB \longrightarrow ijk$:

$$\left(\frac{L_i}{I_i}\right) M_i + 2\left(\frac{L_i}{I_i} + \frac{L_j}{I_j}\right) M_j + \left(\frac{L_j}{I_j}\right) M_k = - \frac{P_j\, a_j\, b_j\, (a_j^2 + 2b_j^2 + 3 a_j b_j)}{I_j\, L_j^2}$$

Here $\quad M_i = M_o \equiv 0$.

$$M_j = M_A$$
$$M_k = M_B$$

$$(0)(0) + 2\left(\frac{0}{\infty} + \frac{10}{I}\right) M_A + \left(\frac{10}{I}\right) M_B = \frac{- 120(4)(6)\left[(4)^2 + 2(6)^2 + 3(4)(6)\right]}{I\,(10)^2}$$

Cancel I from both sides:

$$\boxed{20 M_A + 10 M_B = -4608} \quad\quad\quad (1)$$

Equation (2): $ABC \longrightarrow ijk$:

$$\left(\frac{L_i}{I_i}\right) M_i + 2\left(\frac{L_i}{I_i} + \frac{L_j}{I_j}\right) M_j + \left(\frac{L_j}{I_j}\right) M_k = \frac{-P_i\, a_i\, b_i\, (a_i^2 + 2b_i^2 + 3 a_i b_i)}{I_i\, L_i^2} - \frac{w_j\, L_j^3}{4 I_j}$$

Here, $\quad M_i = M_A$

$$M_j = M_B$$
$$M_k = M_c \equiv 0$$

$$\left(\frac{10}{I}\right)M_A + 2\left(\frac{10}{I} + \frac{10}{I}\right)M_B + \left(\frac{10}{I}\right)(0) = -120\frac{(6)(4)\left[(6)^2 + 2(4)^2 + 3(6)(4)\right]}{I(10)^2} - \frac{50(10)^3}{4I}$$

Cancel I from both sides:

$$\boxed{10M_A + 40M_B = -16532} \qquad \underline{\qquad} (2)$$

Solve equations (1) and (2) simultaneously for M_A and M_B:

$$\begin{array}{r} +1 \\ -2 \end{array} \left| \begin{array}{l} 20M_A + 10M_B = -4608 \\ 10M_A + 40M_B = -16532 \end{array} \right.$$

$$\left.\begin{array}{l} 20M_A + 10M_B = -4608 \\ -20M_A - 80M_B = 33064 \end{array}\right\} \Rightarrow \quad M_B = \underline{\underline{-406.5}} \ \ kN \cdot m$$

$$\therefore \ M_A = \underline{\underline{-27.14}} \ \ kN \cdot m$$

Support Settlement: (Otto Mohr, 1860):

* Consider the two-span section shown in the figure.

* Suppose the supports displace (settle) as shown.

* Take upward displacements to be positive.

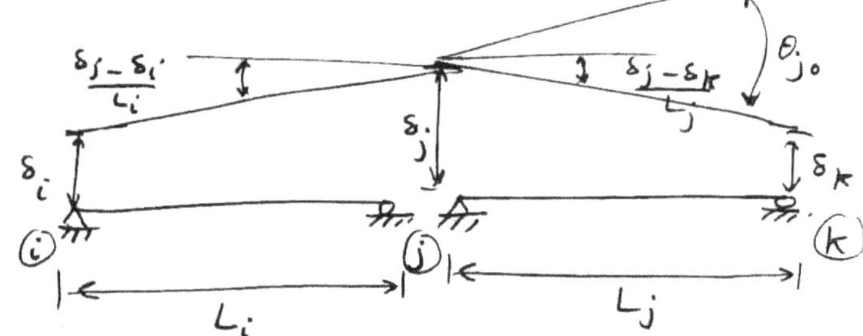

From the geometry shown in the figure:

$$\theta_{jo} = \frac{\delta_j - \delta_i}{L_i} + \frac{\delta_j - \delta_k}{L_j}$$

$$\therefore \ \theta_{jo} = \left[-\frac{\delta_i}{L_i} + \delta_j\left(\frac{1}{L_i} + \frac{1}{L_j}\right) - \frac{\delta_k}{L_j}\right]$$

$\therefore$ the right-hand side of the three-moment equation becomes:

$$-6E\theta_{jo} = \boxed{6E\left[\frac{\delta_i}{L_i} - \delta_j\left(\frac{1}{L_i} + \frac{1}{L_j}\right) + \frac{\delta_k}{L_j}\right]} \ .$$

Example 3 :

Draw the shear force and bending moment diagrams for the continuous beam shown in the figure. There is a 15 mm settlement at support B. Take $EI = 80{,}000\ kN \cdot m^2$.

Solution :

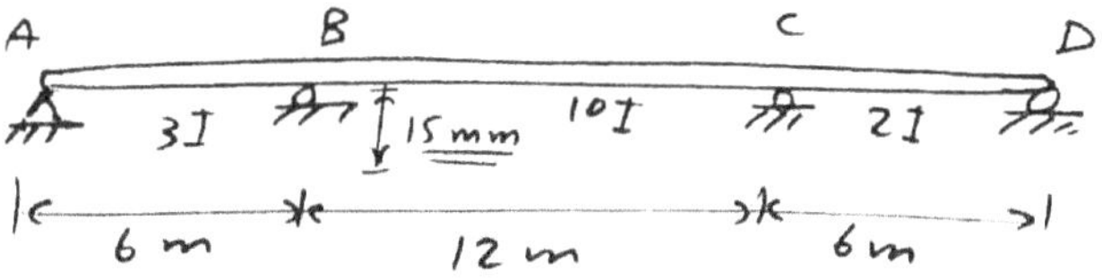

We have two redundants.

$\delta_B = -15\ mm$ (down).

Equation (1): ABC → ijk :

$$\left(\frac{L_i}{I_i}\right)M_i + 2\left(\frac{L_i}{I_i} + \frac{L_j}{I_j}\right)M_j + \left(\frac{L_j}{I_j}\right)M_k = 6E\left[\frac{\delta_i}{L_i} - \delta_j\left(\frac{1}{L_i}+\frac{1}{L_j}\right) + \frac{\delta_k}{L_j}\right]$$

$M_i = 0, \quad M_j = M_B, \quad M_k = M_C.$

$\delta_i = 0, \quad \delta_j = -15\ mm, \quad \delta_k = 0.$

$$\left(\frac{6}{3I}\right)(0) + 2\left(\frac{6}{3I} + \frac{12}{10I}\right)M_B + \left(\frac{12}{10I}\right)M_C = 6E\left[0 - (-15)\left(\frac{1}{6000} + \frac{1}{12000}\right) + 0\right]$$

$$6.4\,M_B + 1.2\,M_C = 6EI\,[0.00375]$$

$$\boxed{6.4\,M_B + 1.2\,M_C = 0.0225\,EI} \qquad (1)$$

Equation (2): BCD → ijk :

$M_i = M_B, \quad M_j = M_C, \quad M_k = 0.$

$\delta_i = -15\ mm, \quad \delta_j = 0, \quad \delta_k = 0.$

$$\left(\frac{12}{10I}\right)M_B + 2\left(\frac{12}{10I} + \frac{6}{2I}\right)M_C + \left(\frac{6}{2I}\right)(0) = 6E\left[\frac{-15}{12000} - (0)\left(\frac{1}{12000} + \frac{1}{6000}\right) + 0\right]$$

$$1.2\,M_B + 8.4\,M_C = 6EI\,[0.00375 - \text{~~0.00125~~}\ (0)]$$

$$\boxed{1.2\,M_B + 8.4\,M_C = -0.0075\,EI} \qquad (2)$$

Solve equations (1) and (2) simultaneously for M_B, M_C :

$$\frac{-8.4}{1.2} \,/\quad 6.4\,M_B + 1.2\,M_C = 0.0225\,EI$$

$$1 \,/\quad 1.2\,M_B + 8.4\,M_C = -0.0075$$

$$-44.8\,M_B - 8.4\,M_C = -0.1575\,EI$$

$$1.2\,M_B + 8.4\,M_C = -0.0075\,EI$$

$$M_B = 0.0037844 \ EI$$

$$= 0.0037844 \ (80,000)$$

$$\therefore \ M_B = \underline{+302.75 \ \text{kN}\cdot\text{m}}$$

$$\therefore \ M_C = \underline{-114.68 \ \text{kN}\cdot\text{m}}$$

Support AB :

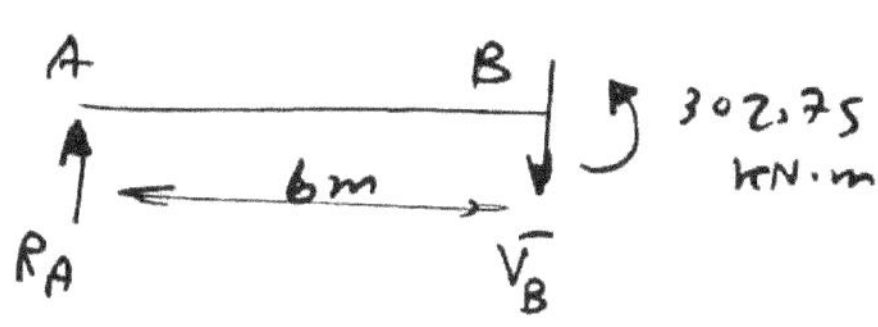

$+\circlearrowleft \ \Sigma M_B = 0:\ \ 302.75 - R_A(6) = 0$

$\qquad R_A = 50.46 \ \text{kN} \uparrow.$

$+\uparrow \ \Sigma F_y = 0:\ \ V_B^- = 50.46 \ \text{kN}.$

Support CD :

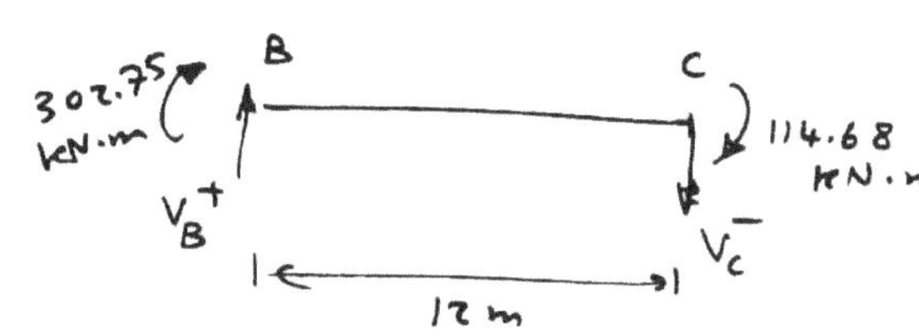

$+\circlearrowleft \ \Sigma M_C = 0:\ \ -114.68 - R_D(6) = 0$

$\qquad R_D = -19.11 \ \text{kN}.$

$\qquad \Rightarrow \ R_D = 19.11 \ \text{kN} \downarrow.$

$+\uparrow \ \Sigma F_y = 0:\ \ V_C^+ = +19.11 \ \text{kN}.$

Support BC :

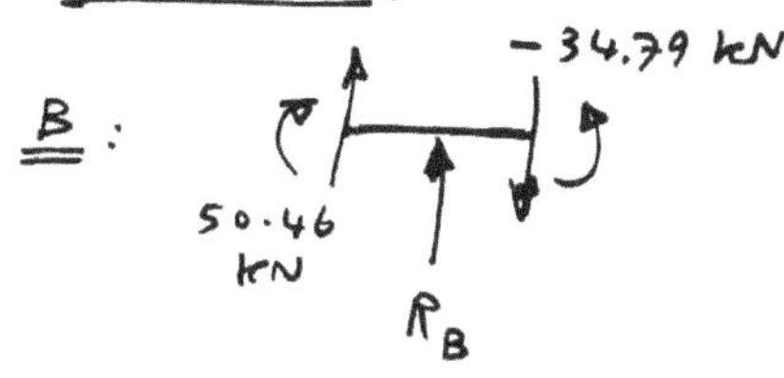

$+\circlearrowleft \ \Sigma M_B = 0:$

$\qquad -302.75 - 114.68 - V_C^-(12) = 0$

$\qquad V_C^- = -34.79 \ \text{kN}.$

$+\uparrow \ \Sigma F_y = 0:\ \ V_B^+ = -34.79 \ \text{kN}.$

Reactions :

B :

$R_B + 50.46 - (-34.79) = 0$

$R_B = -85.25 \ \text{kN}.$

$\Rightarrow \ R_B = 85.25 \ \text{kN} \downarrow.$

C :

$R_C + (-34.79) - 19.11 = 0$

$R_C = 53.9 \ \text{kN} \uparrow.$

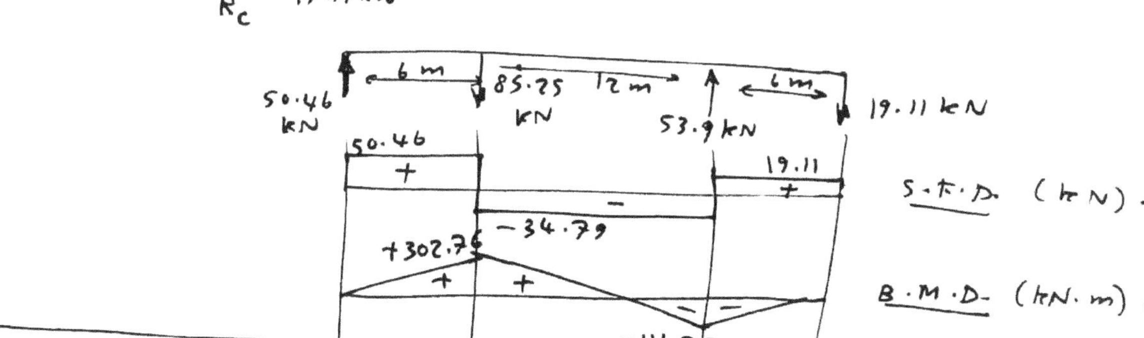

<u>Example 4</u> :

    Find the support moments of the continuous beam shown in the figure.

<u>Solution</u> :

We have three redundants.

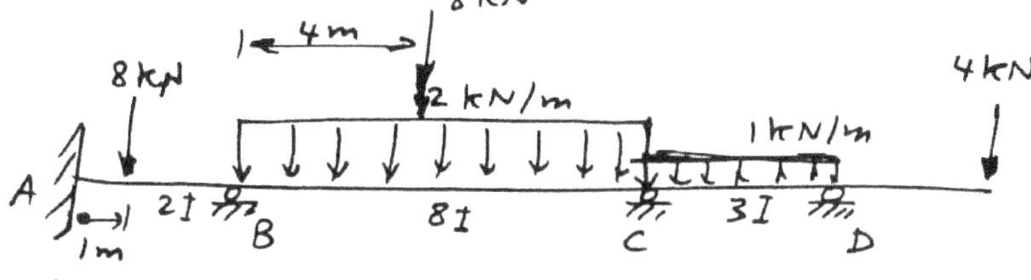

<u>Note</u>: the overhanging part is <u>not</u> considered a span.

$$M_D = -4(2) = -8 \ kN \cdot m.$$

<u>Note</u>: M_D is negative

<u>Note</u>: Use an imaginary span at the fixed end:

$$L = 0$$
$$I = \infty$$

<u>Equation (1)</u>: <u>Span OAB $\longrightarrow$ i j k :</u>

$$\left(\frac{L_i}{I_i}\right)M_i + 2\left(\frac{L_i}{I_i} + \frac{L_j}{I_j}\right)M_j + \left(\frac{L_j}{I_j}\right)M_k = -\frac{P_j \, a_j \, b_j \, (a_j^2 + 2b_j^2 + 3a_j b_j)}{I_j \, L_j^2}$$

$$(0) + 2\left(0 + \frac{4}{2I}\right)M_A + \left(\frac{4}{2I}\right)M_B = -\frac{8\,(1)(3)\left[(1)^2 + 2(3)^2 + 3(1)(3)\right]}{(2I)(4)^2}$$

$$4M_A + 2M_B = -21$$

$$\Longrightarrow \quad \boxed{2M_A + M_B = -10.5} \quad\quad\quad\quad (1)$$

<u>Equation (2)</u>: <u>Span ABC $\longrightarrow$ i j k :</u>

$$\left(\frac{L_i}{I_i}\right)M_i + 2\left(\frac{L_i}{I_i} + \frac{L_j}{I_j}\right)M_j + \left(\frac{L_j}{I_j}\right)M_k = -\frac{P_i \, a_i \, b_i \, (a_i^2 + 2b_i^2 + 3a_i b_i)}{I_i \, L_i^2}$$

$$-\frac{3\,P_j \, L_j^2}{8\,I_j} - \frac{w_j \, L_j^3}{4\,I_j}$$

$$\left(\frac{4}{2I}\right)M_A + 2\left(\frac{4}{2I} + \frac{8}{8I}\right)M_B + \left(\frac{8}{8I}\right)M_C = -\frac{8\,(3)(1)\left[(3)^2 + 2\,(1)^2 + 3\,(3)(1)\right]}{(2I)\,(4)^2}^{(C11)}$$

$$-\frac{3\,(8)\,(8)^2}{8\,(8I)} - \frac{2\,(8)^3}{4(8I)}$$

$$2M_A + 6M_B + M_C = -15 - 24 - 32$$

$$\Longrightarrow \quad \boxed{2M_A + 6M_B + M_C = -71} \quad \underline{\hspace{3cm}} \quad (2)$$

Equation (3) : Span $BCD \longrightarrow ijk$:

$$\left(\frac{L_i}{I_i}\right)M_i + 2\left(\frac{L_i}{I_i} + \frac{L_j}{I_j}\right)M_j + \left(\frac{L_j}{I_j}\right)M_k = -\frac{3P_i L_i^2}{8I_i} - \frac{w_i L_i^3}{4I_i} - \frac{w_j L_j^3}{4I_j}$$

$$\left(\frac{8}{8I}\right)M_B + 2\left(\frac{8}{8I} + \frac{6}{3I}\right)M_C + \frac{6}{3I}(-8) = -\frac{3\,(8)(8)^2}{8\,(8I)} - \frac{2\,(8)^3}{4(8I)} - \frac{1\,(6)^3}{4\,(3I)}$$

$$M_B + 6M_C = -24 - 32 -$$

$$\Longrightarrow \quad \boxed{M_B + 6M_C = -58} \quad \underline{\hspace{3cm}} \quad (3)$$

Solve equations (1), (2) and (3) simultaneously for the three moments:

Write the equations in matrix form:

$$\begin{bmatrix} 2 & 1 & 0 \\ 2 & 6 & 1 \\ 0 & 1 & 6 \end{bmatrix} \begin{Bmatrix} M_A \\ M_B \\ M_C \end{Bmatrix} = \begin{Bmatrix} -10.5 \\ -71 \\ -58 \end{Bmatrix} \quad \text{"matrix equation"}$$

Use program "$\underline{SOLVE}$" to get the results:

$$M_A = +0.002 \ \ kN \cdot m$$

$$M_B = -10.5 \ \ kN \cdot m \ .$$

$$M_C = -7.91 \ \ kN \cdot m$$

Slope–Deflection Method

13.1 Introduction:

* This is a displacement method (the unknowns are displacements).

* The moments at the ends of the members are expressed in terms of the displacements and rotations of these ends.

* The equations are relatively easy to construct regardless of the number of unknowns.

* The slope-deflection method is fairly easy to apply even when the structure becomes relatively complex.

* 1880 – Heinrich Manderla
  1892 – Otto Mohr
  1914 – Alex Bendixen → [named it the Slope-Deflection Method]

  1915 – George Maney (University of Minnesota).

13.2 Derivation of the Slope-Deflection Equation:

* The slope-deflection method is based on the relationship that exists between the moments at the ends of a member (in a rigid frame or continuous beam) and the deformations at the ends.

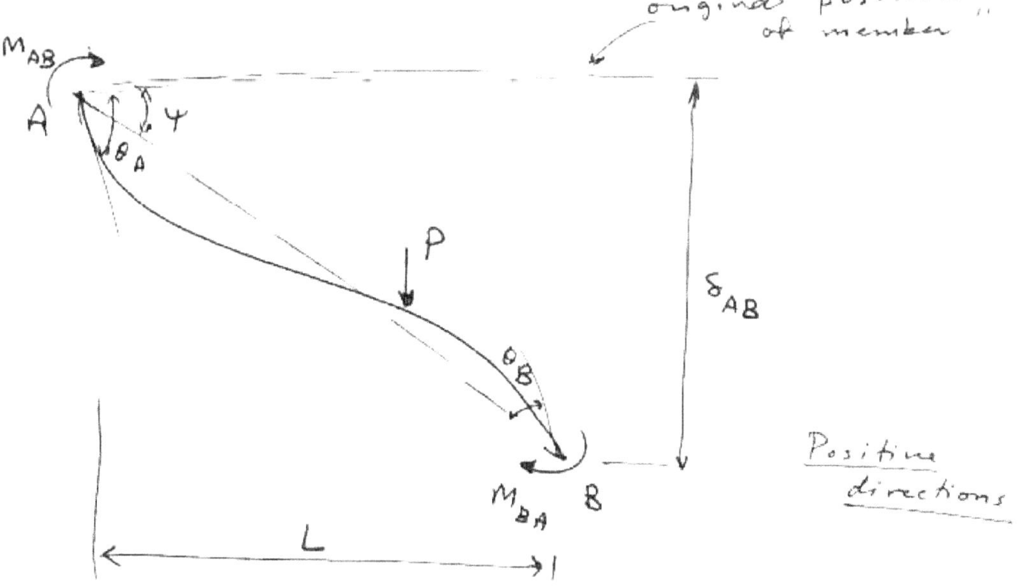

Notation.

(1) Moment M_{AB} acts at end A of member AB.
     Moment M_{BA} acts at end B of member AB

(2) θ_A: rotation at A of member AB } "joint rotation"
     θ_B: rotation at B of member AB }

(3) δ_{AB} : translation of one end of the member relative to the other end in a direction normal to the axis of the member

$$\psi \equiv \psi_{AB} = \frac{\delta_{AB}}{L} \qquad \text{rotation of the axis of the member}$$

Sign Convention :

(1) The moments acting on the ends of the members are positive when clockwise.

(2) The rotations of the ends of the members are also positive when clockwise.

(3) The relative displacement at the ends of a member δ is positive when the axis rotation ψ is clockwise

---

Objective : Derive an expression for the member moments M_{AB} and M_{BA} in terms of the deformations θ_A, θ_B and δ_{AB}, and the load P acting on the member.

* Consider the effect of each variable separately.

---

① End moments due to rotation θ_A :

$$\theta_B = \delta = P = 0$$

$\Rightarrow$ a beam simply-supported at A and fixed at B.

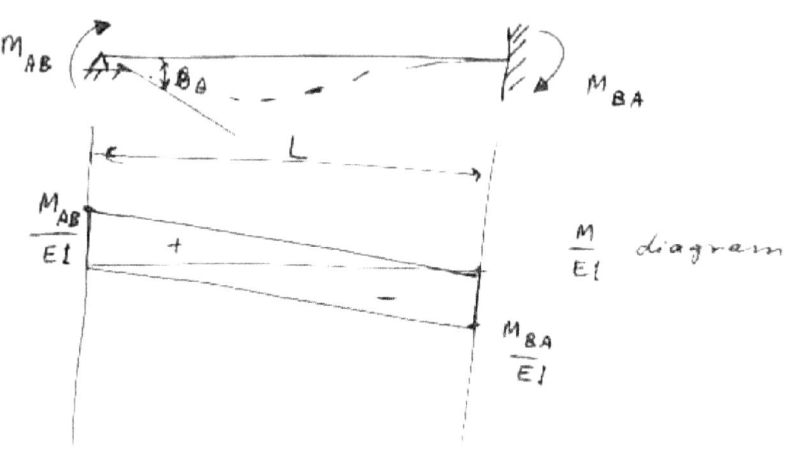

Use the moment-area theorems:

$$\delta_{AB} - 0 = \frac{1}{2}\left(\frac{M_{AB}}{EI}\right)L \cdot \frac{1}{3}L - \frac{1}{2}\left(\frac{M_{BA}}{EI}\right)L \cdot \frac{2}{3}L$$

$$0 = \frac{M_{AB}L^2}{6EI} - \frac{M_{BA}L^2}{3EI}$$

$$\Longrightarrow \quad \boxed{M_{BA} = \frac{1}{2}M_{AB}}$$

$$\theta_A = \frac{1}{2}\left(\frac{M_{AB}}{EI}\right)L - \frac{1}{2}\left(\frac{M_{BA}}{EI}\right)L$$

$$= \frac{(M_{AB} - M_{BA})L}{2EI} .$$

but $\qquad M_{BA} = \frac{1}{2}M_{AB}$

$$\theta_A = \frac{\left(M_{AB} - \frac{1}{2}M_{AB}\right)L}{2EI} = \frac{M_{AB}L}{4EI}$$

$$\Longrightarrow \quad \boxed{M_{AB} = \frac{4EI\,\theta_A}{L}}$$

$$\boxed{M_{BA} = \frac{2EI\,\theta_A}{L}}$$

---

② End moments due to rotation θ_B:

$$\theta_A = \delta = P = 0 .$$

$\Longrightarrow$ a beam simply-supported at B and fixed at A.

$$\boxed{M_{AB} = \frac{2EI\,\theta_B}{L}}$$

$$\boxed{M_{BA} = \frac{4EI\,\theta_B}{L}}$$

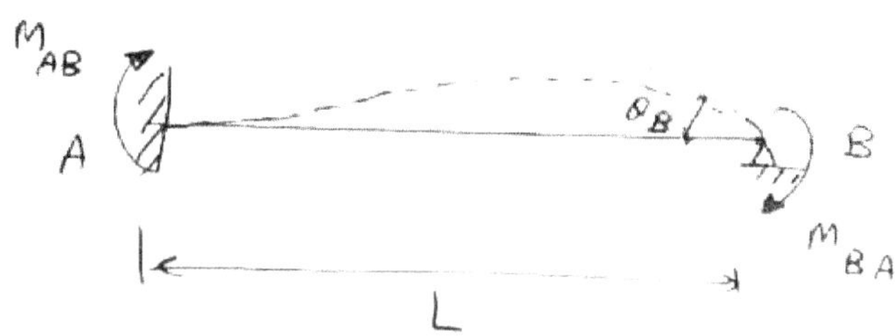

---

③ End moments due to a relative joint displacement δ

$$\theta_A = \theta_B = P = 0 .$$

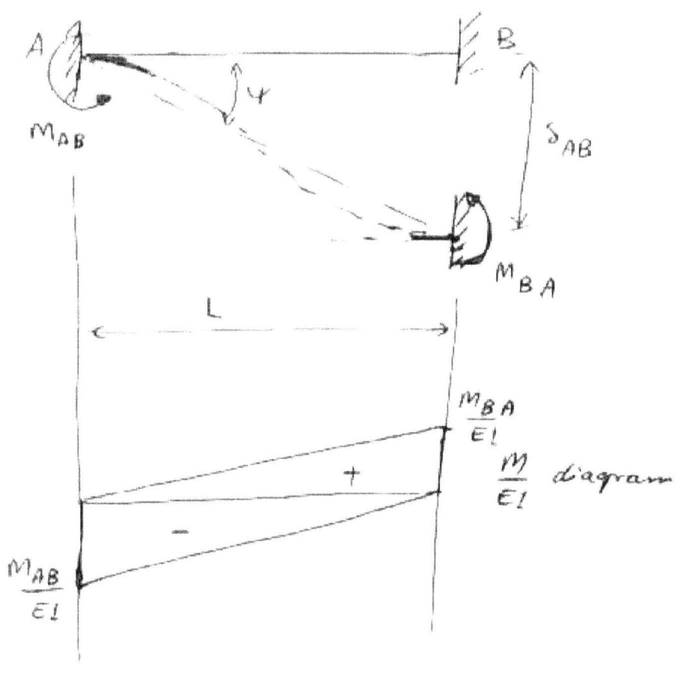

$\dfrac{M_{BA}}{EI}$

$\dfrac{M}{EI}$ diagram

$\dfrac{M_{AB}}{EI}$

$$\Delta\theta_{AB} = 0 \;=\; \tfrac{1}{2}\left(\dfrac{M_{BA}}{EI}\right)L - \tfrac{1}{2}\left(\dfrac{M_{AB}}{EI}\right)L = 0$$

$$\Rightarrow \qquad \boxed{\,M_{AB} = M_{BA}\,}$$

$$S = \delta_{AB} = \tfrac{1}{2}\left(\dfrac{M_{BA}}{EI}\right)\cdot L\left(\tfrac{1}{3}L\right) - \tfrac{1}{2}\left(\dfrac{M_{AB}}{EI}\right)\cdot L\left(\tfrac{2}{3}L\right)$$

$$S = \dfrac{M_{BA}\,L^2}{6EI} - \dfrac{M_{AB}\,L^2}{3EI} \qquad , \text{ but } \quad M_{AB} = M_{BA}$$

$$S = \dfrac{-M_{AB}\,L^2}{6EI} = -6\,\dfrac{M_{BA}\,L^2}{6EI}$$

$$\Rightarrow \qquad \boxed{\,M_{AB} = M_{BA} = \dfrac{-6\,EI\,S}{L^2}\,}$$

Note
Positive translation leads to negative end moments.

$\textcircled{4}$ End moments due to loads acting on the member.

$$\theta_A = \theta_B = S = 0$$

$\Rightarrow$ fixed-end beam.

$\left.\begin{array}{l} M_{F_{AB}} \\[2mm] M_{F_{BA}} \end{array}\right\}$ "Fixed-end moments"

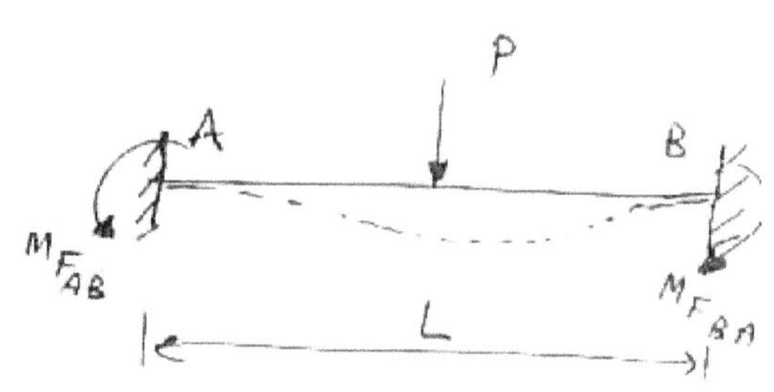

* Fixed End Moments depend on the loads on the structure. [34

Examples:

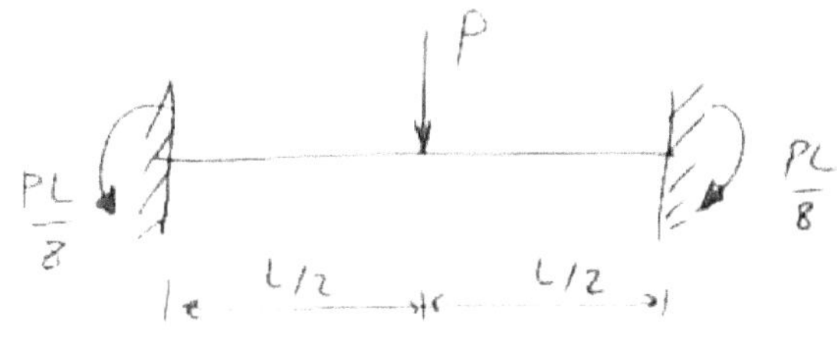

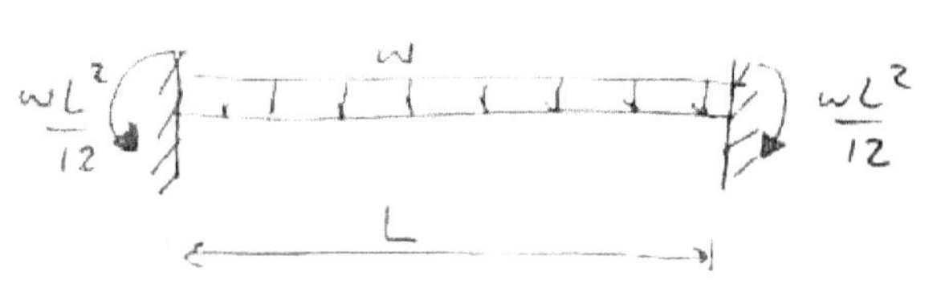

Combining the above four cases, we get:

$$M_{AB} = \frac{4EI\theta_A}{L} + \frac{2EI\theta_B}{L} - \frac{6EI\delta}{L^2} \pm M_{F_{AB}}$$

$$\Rightarrow \boxed{M_{AB} = \frac{2EI}{L}\left(2\theta_A + \theta_B - 3\frac{\delta}{L}\right) \pm M_{F_{AB}}}$$

Similarly,

$$\boxed{M_{BA} = \frac{2EI}{L}\left(2\theta_B + \theta_A - 3\frac{\delta}{L}\right) \pm M_{F_{BA}}}$$

Slope-Deflection Equations

* Note: The slope-deflection equations consider only bending deformations.

Deformations due to shear and axial loads in bending members are ignored.

13.3 Alternate Derivation of Slope-Deflection Equation:

* Use the differential equation.

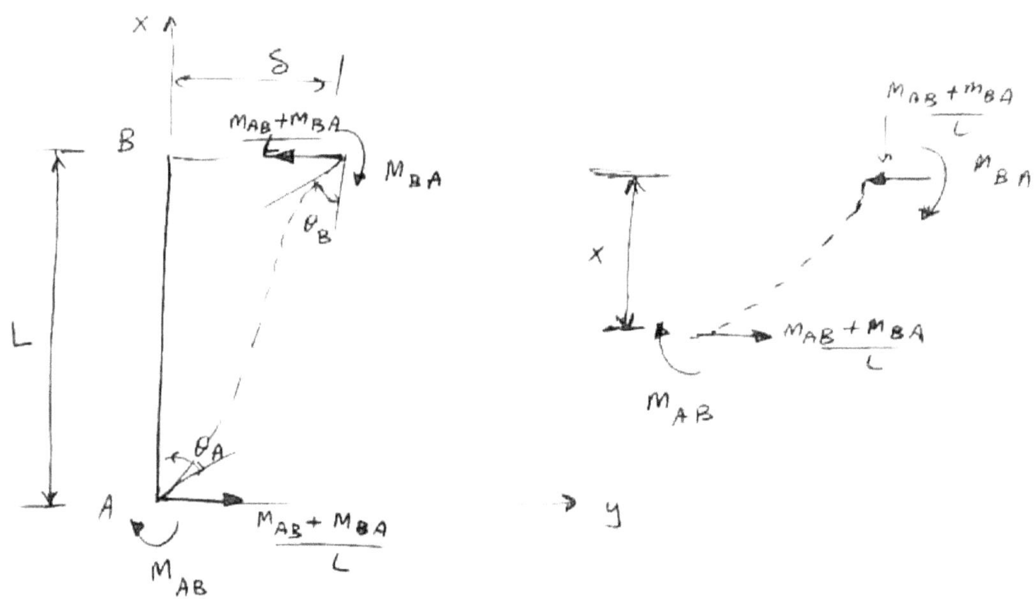

$$EI\,y'' = \frac{M_{AB}}{L}(x-L) + M_{BA}\frac{x}{L} \qquad \text{(differential equation)}.$$

Integrate:

$$EI\,y' = \frac{M_{AB}}{L}\left(\frac{x^2}{2}-Lx\right) + M_{BA}\frac{x^2}{2L} + C_1$$

$$EI\,y = \frac{M_{AB}}{L}\left(\frac{x^3}{6}-\frac{Lx^2}{2}\right) + M_{BA}\frac{x^3}{6L} + C_1 x + C_2$$

At A: $y = 0$ and $y' = \theta_A$ $(x=0)$:

$$\Rightarrow \quad C_2 = 0 \quad \text{and} \quad EI\theta_A = C_1$$

At B: $y' = \theta_B$ and $y = \delta$ $(x=L)$:

$$\frac{M_{AB}}{L}\left(\frac{L^2}{2}-L^2\right) + M_{BA}\frac{L^2}{2L} + EI\theta_A = EI\theta_B \qquad \text{---- (1)}$$

$$\frac{M_{AB}}{L}\left(\frac{L^3}{6}-\frac{L^3}{2}\right) + M_{BA}\frac{L^3}{6L} + EI\theta_A\cdot L + 0 = EI\delta \qquad \text{---- (2)}$$

$$-\tfrac{2}{3}L \;\Big/\; -\tfrac{1}{2}M_{AB}L + \tfrac{1}{2}M_{BA}L + EI\theta_A = EI\theta_B$$

$$1 \;\Big/\; -\tfrac{1}{3}M_{AB}L^2 + \tfrac{1}{6}M_{BA}L^2 + EIL\theta_A = EI\delta$$

$$\tfrac{1}{3} M_{AB} L^2 - \tfrac{1}{3} M_{BA} L^2 - \tfrac{2}{3} EIL\theta_A - \ - \tfrac{2}{3} EIL\theta_B$$

$$-\tfrac{1}{3} M_{AB} L^2 + \tfrac{1}{6} M_{BA} L^2 + EIL\theta_A = EI\delta$$

$$\left(\tfrac{1}{6} - \tfrac{1}{3}\right) M_{BA} L^2 + \left(1 - \tfrac{2}{3}\right) EIL\theta_A - EI\delta - \tfrac{2}{3} EIL\theta_B$$

$$-\tfrac{1}{6} M_{BA} L + \tfrac{1}{3} EI\theta_A = EI\tfrac{\delta}{L} - \tfrac{2}{3} EI\theta_B$$

$$\tfrac{1}{L} M_{BA} L = \tfrac{2}{3} EI\theta_B + \tfrac{1}{3} EI\theta_A - EI\tfrac{\delta}{L}$$

$$M_{BA} = 4\frac{EI}{L}\theta_B + 2\frac{EI}{L}\theta_A - 6\frac{EI\delta}{L^2}$$

$$\Longrightarrow \quad M_{BA} = \frac{2EI}{L}\left(2\theta_B + \theta_A - 3\frac{\delta}{L}\right) \quad + M_{FBA} \quad \Big\} \text{ for the loads}$$

$$\Longrightarrow \quad M_{AB} = \frac{2EI}{L}\left(2\theta_A + \theta_B - 3\frac{\delta}{L}\right) \quad + M_{FAB}$$

13.4 Application of the Slope-Deflection Method:

* The slope-deflection method employs an <u>indirect</u> approach to determine the reaction and internal moments.

* The unknown moments are written in terms of the unknown joint displacements using the slope-deflection equations

* Equations of Equilibrium are then written to determine the unknown joint displacements.

* The values of the joint displacements are substituted back into the slope-deflection equations.

* The slope-deflection method is an "equilibrium method"

Example 13·1 :

Determine the reactions and draw the moment diagram for the beam shown in the figure.

Solution :

Joint rotations and translations are the unknowns.

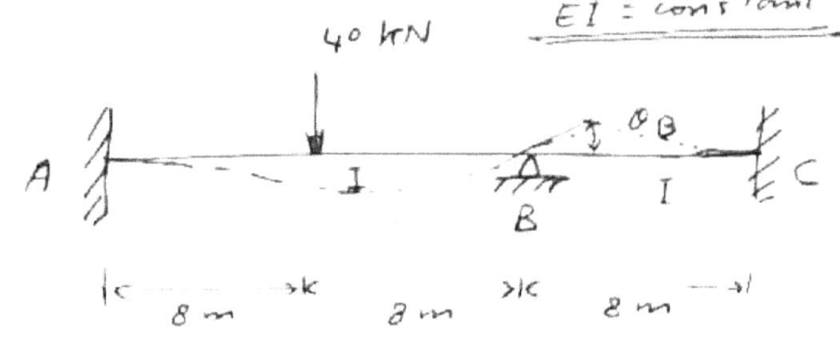

Degree of kinematic indeterminacy = 1   (rotation at B, i.e. θ_B)

∴ Number of degrees of freedom = 1.

∴ I've have **one** unknown, θ_B.

* Write moment equilibrium equations at the joints that are free to rotate.

Step 1. Equilibrium Equations.

Joint B :

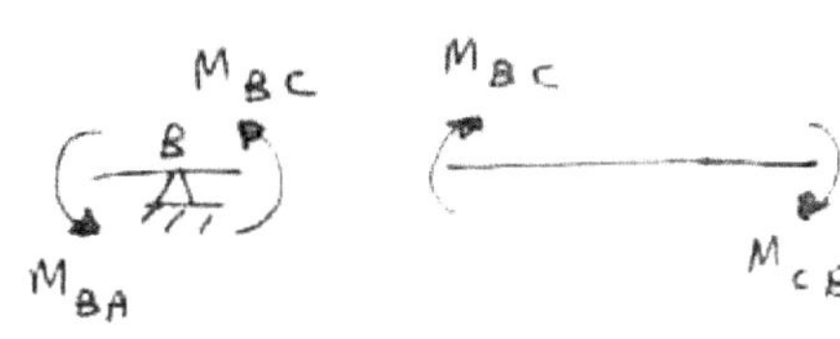

$+\circlearrowright\,\Sigma M_B = 0$:

$$\boxed{M_{BA} + M_{BC} = 0}$$

* Assume all member moments to be positive (clockwise).

Step 2 : Slope-Deflection Equations:

First, evaluate the fixed-end moments :

$$M_{F_{AB}} = -\frac{PL}{8} = \frac{-40(16)}{8} = -80 \quad kN\cdot m$$

$$M_{F_{BA}} = +\frac{PL}{8} = +80 \quad kN\cdot m$$

$$M_{F_{BC}} = 0$$

$$M_{F_{CB}} = 0$$

Note : $\theta_A = 0, \quad \theta_C = 0, \quad \delta_{AB} = 0, \quad \delta_{BC} = 0$

$$M_{AB} = 2\frac{EI}{L}\left(2\theta_A + \theta_B - 3\frac{\delta_{AB}}{L}\right) \pm M_{F_{AB}}$$

$$= \frac{2EI}{16}\left(0 + \theta_B - 0\right) - 80$$

$$\Longrightarrow M_{AB} = \frac{EI}{8}\theta_B - 80 \quad\text{———————— (1)}$$

$$M_{BA} = \frac{2EI}{L}\left(2\theta_B + \theta_A - 3\frac{\delta_{AB}}{L}\right) \pm M_{F_{BA}}$$

$$= \frac{2EI}{16}\left(2\theta_B + 0 - 0\right) + 80$$

$$\Longrightarrow M_{BA} = \frac{EI}{4}\theta_B + 80 \quad\text{———————— (2)}$$

$$M_{BC} = \frac{2EI}{L}\left(2\theta_B + \theta_C - 3\frac{\delta_{BC}}{L}\right) \pm M_{F_{BC}}$$

$$= \frac{2EI}{8}\left(2\theta_B + 0 - 0\right) + 0$$

$$\Longrightarrow M_{BC} = \frac{EI}{2}\theta_B \quad\text{———————— (3)}$$

$$M_{CB} = \frac{2EI}{L}\left(2\theta_C + \theta_B - 3\frac{\delta_{BC}}{L}\right) \pm M_{F_{CB}}$$

$$= \frac{2EI}{8}\left(0 + \theta_B - 0\right) \pm M_{F_{CB}}$$

$$\Longrightarrow M_{CB} = \frac{EI}{4}\theta_B \quad\text{———————— (4)}$$

Step 3: Substitute in the equilibrium equations:

$$M_{BA} + M_{BC} = 0$$

$$\boxed{\left(\frac{EI}{4}\theta_B + 80\right) + \left(\frac{EI}{2}\theta_B\right) = 0} \quad \underline{\text{one}} \text{ equation only.}$$

$$\frac{3}{4}EI\theta_B = -80$$

$$\Longrightarrow EI\theta_B = -\frac{320}{3} \Longrightarrow \boxed{EI\theta_B = -106.7}$$

Step 4: Determine the member moments:

$$M_{AB} = \frac{1}{8}EI\theta_B - 80 = \frac{1}{8}(-106.7) - 80 = -93.3 \text{ kN}\cdot$$

$$M_{BA} = \frac{1}{4}EI\theta_B + 80 = \frac{1}{4}(-106.7) + 80 = +53.3 \text{ kN}\cdot$$

$$M_{BC} = \frac{?}{?} EI\,\theta_B = \frac{1}{2}(-106.7) = -53.3 \ kN\cdot m$$

$$M_{CB} = \frac{1}{?} EI\,\theta_B = \frac{1}{4}(-106.7) = -26.7 \ kN\cdot m$$

Step 5   Reactions:

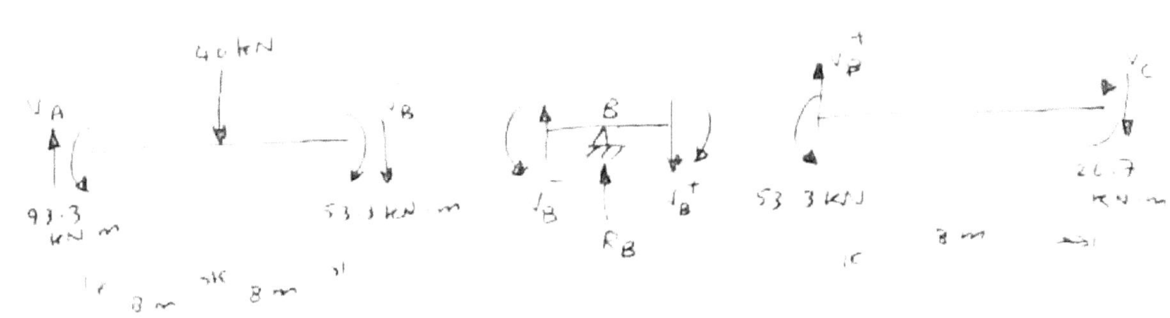

Member AB $\curvearrowright \ \Sigma M_A = 0$ $93.3 - 53.3 - \bar{V}_B(16) - 40(2) = 0$

$$\Rightarrow \bar{V}_B = -17.5 \ kN$$

$$\uparrow \ \Sigma F_y = 0 \quad V_A - 40 - (-17.5) = 0$$

$$\Rightarrow V_A = 27.5 \ kN$$

Member BC $\curvearrowright \ \Sigma M_B = 0:$ $53.3 + 26.7 - V_C(8) = 0$

$$\Rightarrow V_C = 10 \ kN$$

$$\uparrow \ \Sigma F_y = 0: \quad V_B^+ - (10) = 0 \Rightarrow V_B^+ = 10 \ kN$$

Support B $\uparrow \ \Sigma F_y = 0:$ $R_B + (-17.5) - (10) = 0$

$$\Rightarrow R_B = 27.5 \ kN$$

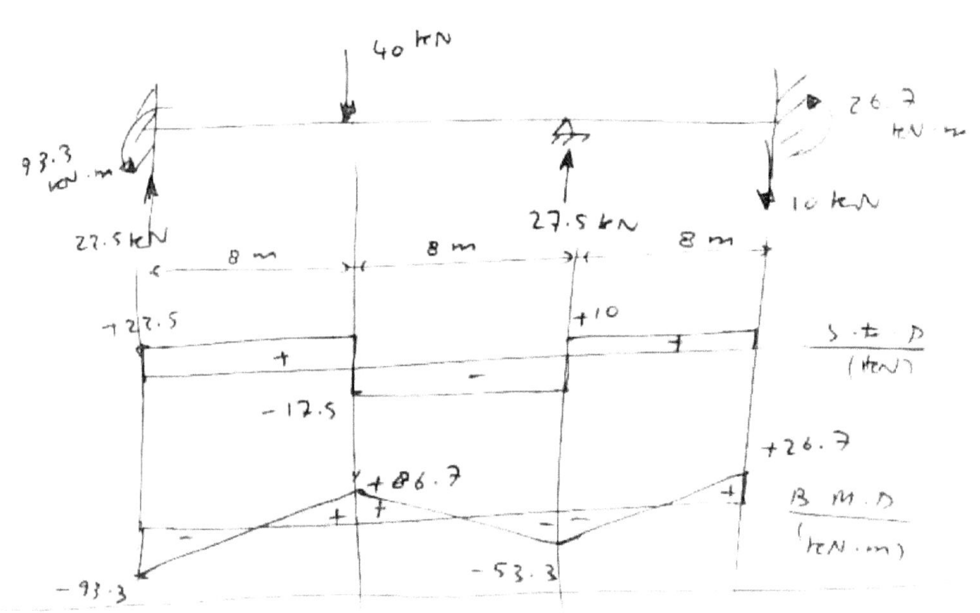

Example 13.2

Determine the reactions and draw the bending moment diagram for the beam shown in the figure

Solution:

(1) Unknowns:

θ_A, θ_B, θ_C, θ_D

(2) Equations of Equilibrium (written at the joints that are free to rotate)

At joint A: $M_{AB} = 0$ ... (1)

At joint B: $M_{BA} + M_{BC} = 0$ ... (2)

At joint C: $M_{CB} + M_{CD} = 0$ ... (3)

At joint D: $M_{DC} = 0$ ... (4)

(3) Fixed-End Moments:

$$M_{F_{AB}} = -\frac{Pab^2}{L^2} = -\frac{20(8)(12)^2}{(20)^2} = -57.6 \ k\text{-}ft$$

$$M_{F_{BA}} = +\frac{Pba^2}{L^2} = +\frac{20(12)(8)^2}{(20)^2} = +38.4 \ k\text{-}ft$$

$$M_{F_{BC}} = -\frac{wL^2}{12} = -\frac{2(20)^2}{12} = -66.67 \ k\text{-}ft$$

$$M_{F_{CB}} = +\frac{wL^2}{12} = +66.67 \ k\text{-}ft$$

$$M_{F_{CD}} = 0$$

$$M_{F_{DC}} = 0$$

(4) Slope-Deflection Equations:

$$M_{AB} = \frac{2EI}{L}\left(2\theta_A + \theta_B - 3\frac{\Delta_{AB}}{L}\right) + M_{F_{AB}}$$

$$\frac{2EI}{20}\left(2\theta_A + \theta_B - 3(0)\right) - 57.6$$

$$0.2EI\theta_A + 0.1EI\theta_B - 57.6$$

$$\Rightarrow M_{AB} =$$

$$M_{BA} = \frac{2EI}{L}\left(2\theta_B + \theta_A - 0\right) + M_{F_{BA}}$$

$$0.2EI\theta_B + 0.1EI\theta_A + 38.4$$

$$\Rightarrow M_{BA} =$$

$$M_{BC} = \frac{2EI}{L}(2\theta_B + \theta_C - \omega) \pm M_{F\,BC}$$

$$= \frac{2EI}{20}(2\theta_B + \theta_C) - 66.67$$

$$= 0.2EI\theta_B + 0.1EI\theta_C - 66.67$$

$$M_{CB} = 0.2EI\theta_C + 0.1EI\theta_B + 66.67$$

$$M_{CD} = 0.2EI\theta_C + 0.1EI\theta_D$$

$$M_{DC} = 0.2EI\theta_D + 0.1EI\theta_C$$

---

(5) Substitute (4) into (2) and solve the system.

$$M_{AB} = 0$$

$$\Rightarrow \quad 0.2EI\theta_A + 0.1EI\theta_B = 57.6 \qquad \text{——— (1)}$$

$$M_{BA} + M_{BC} = 0$$

$$0.2EI\theta_B + 0.1EI\theta_A + 38.4 + 0.2EI\theta_B + 0.1EI\theta_C - 66.67 = 0$$

$$0.1EI\theta_A + 0.4EI\theta_B + 0.1EI\theta_C = 28.27 \qquad \text{——— (2)}$$

$$M_{CB} + M_{CD} = 0$$

$$\Rightarrow \quad 0.2EI\theta_C + 0.1EI\theta_B + 66.67 + 0.2EI\theta_C + 0.1EI\theta_D = 0$$

$$0.1EI\theta_B + 0.4EI\theta_C + 0.1EI\theta_D = -66.67 \qquad \text{——— (3)}$$

$$M_{DC} = 0$$

$$\Rightarrow \quad 0.2EI\theta_D + 0.1EI\theta_C = 0$$

$$0.1EI\theta_C + 0.2EI\theta_D = 0 \qquad \text{——— (4)}$$

Write the equations in matrix form:

$$EI\begin{bmatrix} 0.2 & 0.1 & 0 & 0 \\ 0.1 & 0.4 & 0.1 & 0 \\ 0 & 0.1 & 0.4 & 0.1 \\ 0 & 0 & 0.1 & 0.2 \end{bmatrix}\begin{Bmatrix} \theta_A \\ \theta_B \\ \theta_C \\ \theta_D \end{Bmatrix} = \begin{Bmatrix} 57.6 \\ 28.27 \\ -66.67 \\ 0 \end{Bmatrix}$$

Use program <u>SOLVE</u> to get

$$EI\theta_A = 259.2$$
$$EI\theta_B = 57.6$$
$$EI\theta_C = -206.9$$
$$EI\theta_D = 103.5$$

(6) Back-substitute.

$$M_{AB} = 0.2(259.2) + 0.1(57.6) - 57.6 = 0$$
$$M_{BA} = 0.2(57.6) + 0.1(259.2) + 38.4 = +75.2 \ k-ft.$$
$$M_{BC} = 0.2(57.6) + 0.1(-206.9) - 66.67 = -75.24 \ k-ft$$
$$M_{CB} = 0.2(-206.9) + 0.1(57.6) + 66.67 = +31.05 \ k-ft.$$
$$M_{CD} = 0.2(-206.9) + 0.1(103.5) = -31.03 \ k-ft$$
$$M_{DC} = 0$$

(7) Find the <u>reactions</u>:

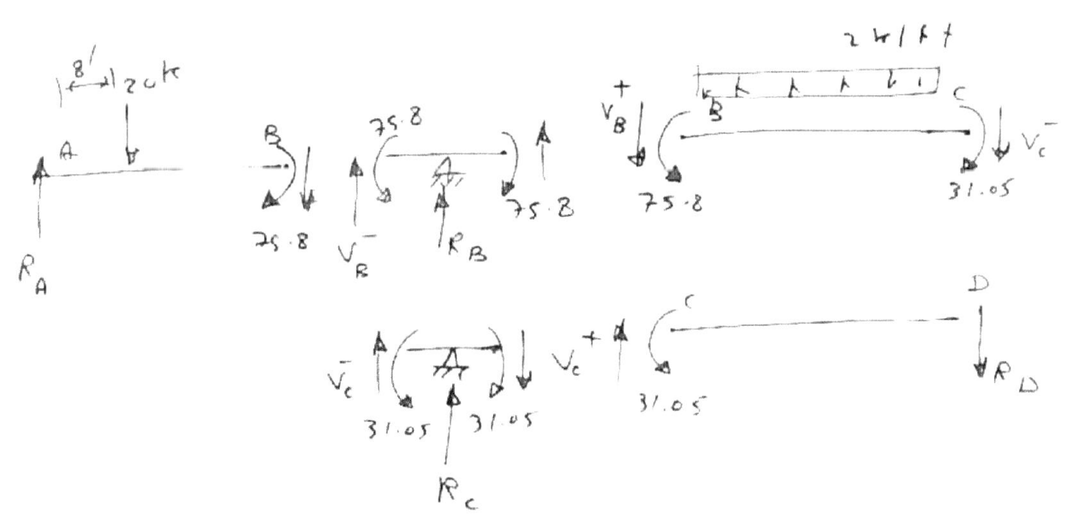

$\underline{AB}$: $+\circlearrowleft \Sigma M_A = 0$: $-V_B^-(20) - 75.8 - 20(8) = 0$

$\Longrightarrow V_B^- = -11.79 \ k.$

$+\uparrow \Sigma F_y = 0$: $R_A - 20 - (-11.79) = 0$

$R_A = 8.21 \ k \uparrow$.

$\underline{BC}$:

$+\circlearrowleft \Sigma M_B = 0$: $-V_C^-(20) - 31.05 - 2(20)(10) \ 75.8 = 0$

$\Longrightarrow V_C^- = -17.76 \ k.$

$+\uparrow \Sigma F_y = 0$: $-V_B^+ - (-17.76) - 2(20) = 0 \Longrightarrow V_B^+ = -22.24 \ k.$

(CD) $+\circlearrowleft \Sigma M_c = 0$: $-R_D(20) + 31.05 = 0$

$\Rightarrow R_D = 1.55\ k \downarrow$

$+\uparrow \Sigma F_y = 0$: $V_c^+ - 1.55 = 0 \Rightarrow V_c^+ = 1.55\ k$

Joint B : $R_B + (-11.79) + (-22.24) = 0$

$\Rightarrow R_B = 34.03\ k \uparrow$

Joint C : $R_c + (-17.76) - (1.55) = 0$

$\Rightarrow R_c = 19.31\ k \uparrow$

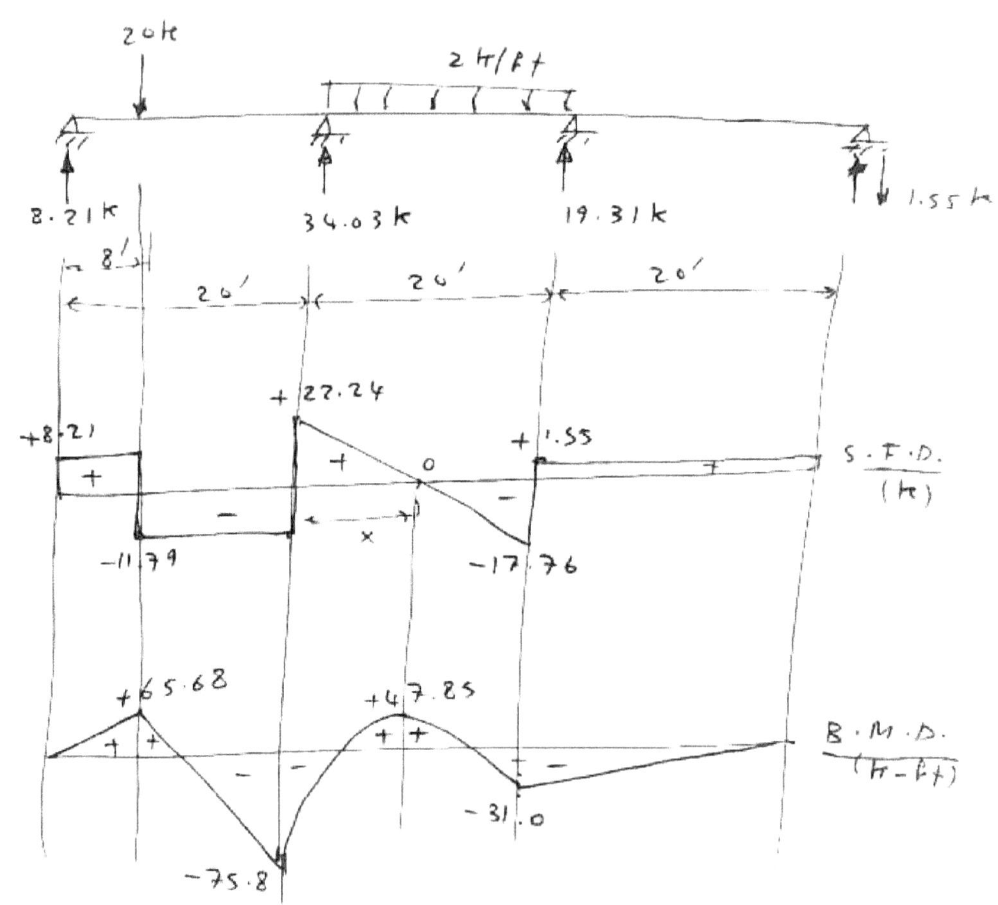

Locate point O of zero shear.

$$\frac{x}{22.24} = \frac{20-x}{17.76}$$

$$17.76x = 20(22.24) - 22.24x$$

$$\Rightarrow x = 11.12\ ft$$

$$M_o = -75.8 + \tfrac{1}{2}(22.24)(11.12) = +47.85\ k\text{-}ft$$

Example 13.3 :

Determine the reactions and draw the bending-moment diagram for the structure shown in the figure.

Solution :

(1) Unknowns :

θ_B, θ_C .

(2) Equations of Equilibrium :

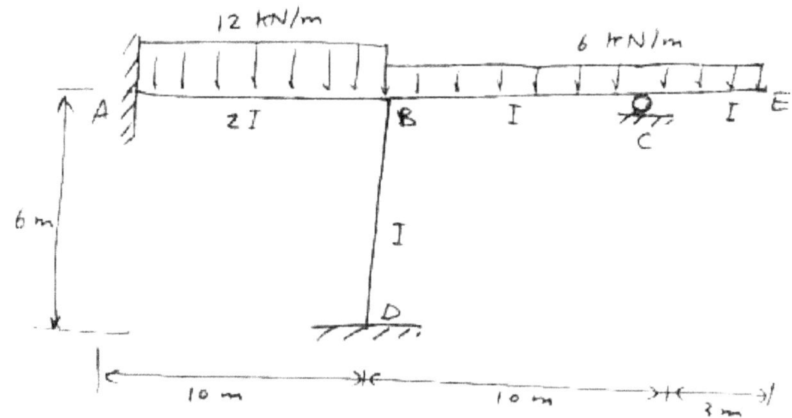

At joint B : $M_{BA} + M_{BC} + M_{BD} = 0$ —— (1)

At joint C : $M_{CB} - 27 = 0$ —————— (2)

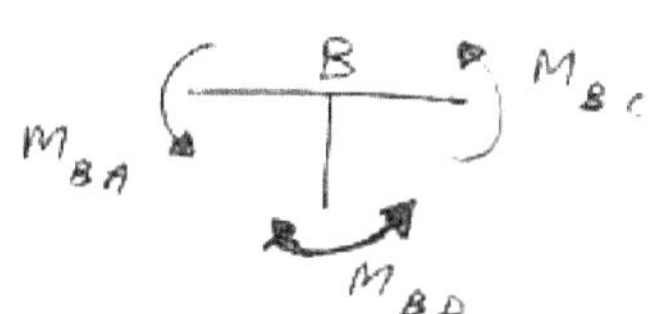

(3) Fixed-End Moments .

$$M_{F_{AB}} = -\frac{w\ell^2}{12} = -\frac{12(10)^2}{12} = -100 \ \text{kN·m}$$

$$M_{F_{BA}} = +100 \ \text{kN·m}$$

$$M_{F_{BC}} = -\frac{w\ell^2}{12} = -\frac{6(10)^2}{12} = -50 \ \text{kN·m}$$

$$M_{F_{CB}} = +50 \ \text{kN·m}$$

$$M_{F_{BD}} = 0$$

$$M_{F_{DB}} = 0$$

$$\frac{6(3)\left(\frac{3}{2}\right)}{} = 27 \ \text{kN·m}$$

(4) Slope-Deflection Equations .

$$M_{AB} = \frac{2EI}{L}\left(2\theta_A + \theta_B - 3\frac{\delta_{AB}}{L}\right) + M^F_{AB}$$

$$= \frac{2E(2I)}{10}\left(0 + \theta_B - 0\right) - 100$$

$$\Rightarrow \quad M_{AB} = 0.4EI\theta_B - 100$$

$$M_{BA} = \frac{2EI}{L}\left(2\theta_B + \theta_A - 3\frac{\delta_{AB}}{L}\right) + M^F_{BA}$$

$$= \frac{2E(2I)}{10}\left(2\theta_B + 0 - 0\right) + 100$$

$$\Rightarrow \quad M_{BA} = 0.8 EI\theta_B + 100$$

$$M_{BC} = \frac{2EI}{L}\left(2\theta_B + \theta_C - 3\frac{\delta_{BC}}{L}\right) + m^F_{BC}$$

$$= \frac{2EI}{10}\left(2\theta_B + \theta_C - 0\right) + (-50)$$

$$\Rightarrow \quad M_{BC} = 0.4 EI\theta_B + 0.2 EI\theta_C - 50$$

$$M_{CB} = \frac{2EI}{10}\left(2\theta_C + \theta_B - 0\right) + 50$$

$$\Rightarrow \quad M_{CB} = 0.4 EI\theta_C + 0.2 EI\theta_B + 50$$

$$M_{BD} = \frac{2EI}{6}\left(2\theta_B + \cancel{\theta_D}^{0} - 0\right) + 0$$

$$\Rightarrow \quad M_{BD} = 0.667 EI\theta_B$$

$$M_{DB} = \frac{2EI}{6}\left(\cancel{2\theta_D}^{0} + \theta_B - 0\right) + 0$$

$$\Rightarrow \quad M_{DB} = 0.333 EI\theta_B$$

(5) Substitute (4) into (2):

Equation (2) $\Rightarrow \quad M_{CB} - 27 = 0$

$$0.4 EI\theta_C + 0.2 EI\theta_B + 50 - 27 = 0$$

$$\Rightarrow \quad \boxed{0.2 EI\theta_B + 0.4 EI\theta_C = -23} \quad\text{——— (2)}$$

Equation (1) $\Rightarrow$

$$M_{BA} + M_{BC} + M_{BD} = 0$$

$$(0.8 EI\theta_B + 100) + (0.4 EI\theta_B + 0.2 EI\theta_C - 50) + (0.667 EI\theta_B) = 0$$

$$\boxed{1.867 EI\theta_B + 0.2 EI\theta_C = -50} \quad\text{——— (1)}$$

Solve equations (2) & (1) simultaneously:

multiply (1) by -0.5 :

$$-0.1 EI\theta_B - 0.2 EI\theta_C = +11.5$$

$$1.867 EI\theta_B + 0.2 EI\theta_C = -50$$

$$\overline{\rule{0pt}{1em}\quad\quad\quad\quad\quad\quad\quad}$$

$$1.767 EI\theta_B = -38.5$$

$$\Rightarrow \quad \boxed{EI\theta_B = -21.8}$$

$$\Rightarrow \quad \boxed{EI\theta_C = -46.6}$$

(6) Get the member moments:

$$M_{AB} = 0.4(-21.8) - 100 = -108.72 \text{ kN·m}$$

$$M_{BA} = 0.8(-21.8) + 100 = +82.56 \text{ kN·m}$$

$$M_{BC} = 0.4(-21.8) + 0.2(-46.6) - 50 = -68.04 \text{ kN·m}$$

$$M_{CB} = 0.4(-46.6) + 0.2(-21.8) + 50 = +27.0 \text{ kN·m} \quad \text{(already known).}$$

$$M_{BD} = 0.667(-21.8) = -14.54 \text{ kN·m}$$

$$M_{DB} = 0.333(-21.8) = -7.27 \text{ kN·m}$$

(7) Determine the reactions:

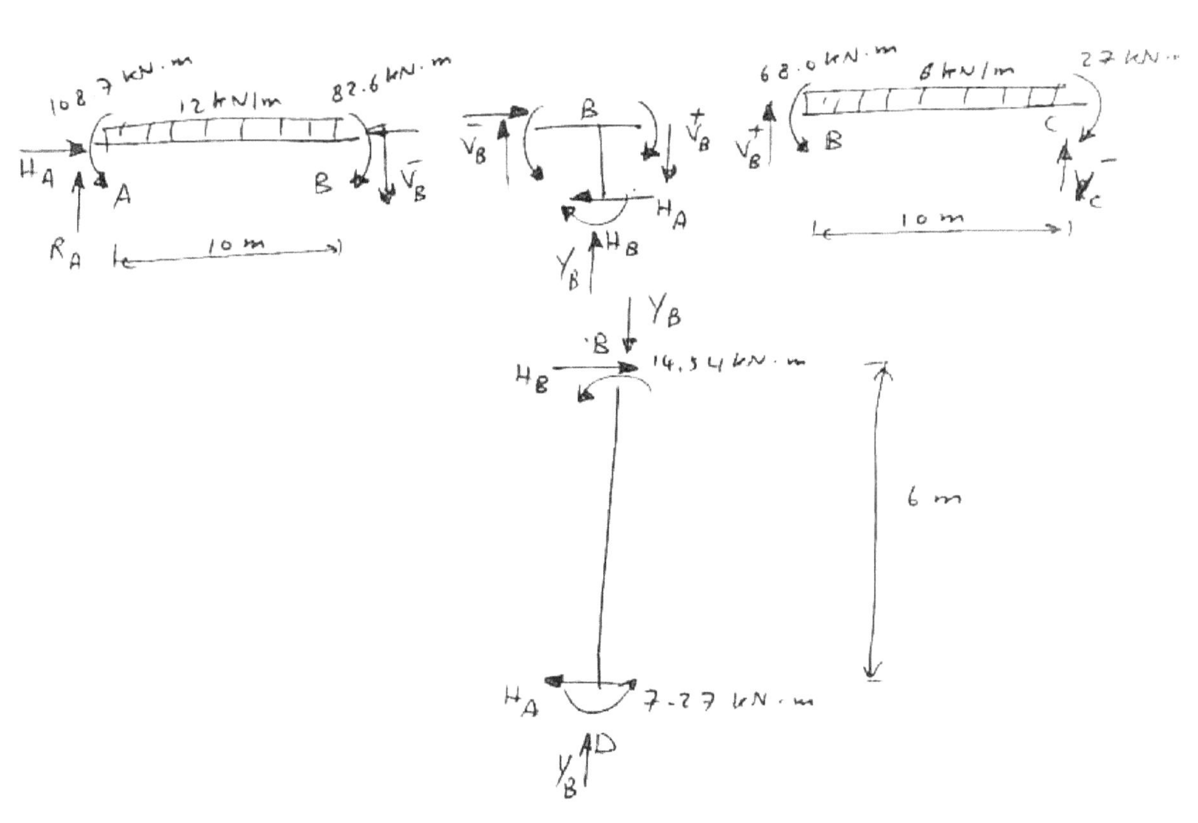

$\underline{AB}: \quad +\circlearrowleft \Sigma M_A = 0: \quad -\bar{V}_B(10) - 82.6 + 108.7 - 12(10)(5) = 0$

$$\bar{V}_B = -57.4 \text{ kN}$$

$+\uparrow \Sigma F_y = 0: \quad R_A - (-57.4) - 12(10) = 0$

$$R_A = 62.6 \text{ kN} \uparrow$$

$\underline{BC}: \quad +\circlearrowleft \Sigma M_B = 0: \quad 68.0 - 27 + \bar{V}_C(10) - 6(10)(5) = 0$

$$\bar{V}_C = 25.9 \text{ kN}$$

$$V_C^+ = 6(3) = 18 \text{ kN}$$

Joint C: $\quad R_C = 25.9 + 18 = 43.9 \text{ kN}$

$$+\uparrow\Sigma F_y = 0: \quad V_B^+ + (25.9) - 6(10) = 0$$
$$V_B^+ = 34.1 \text{ kN}$$

$\underline{\text{Joint } B}: \quad Y_B + V_B^- - V_B^+ = 0$
$$Y_B + (-57.4) - (34.1) = 0$$
$$Y_B = 91.5 \text{ kN}$$

$\underline{BD}: \quad +\circlearrowright\Sigma M_B = 0: \quad 14.54 + 7.27 - H_A(6) = 0$
$$H_A = 3.64 \text{ kN}$$

Member BD:

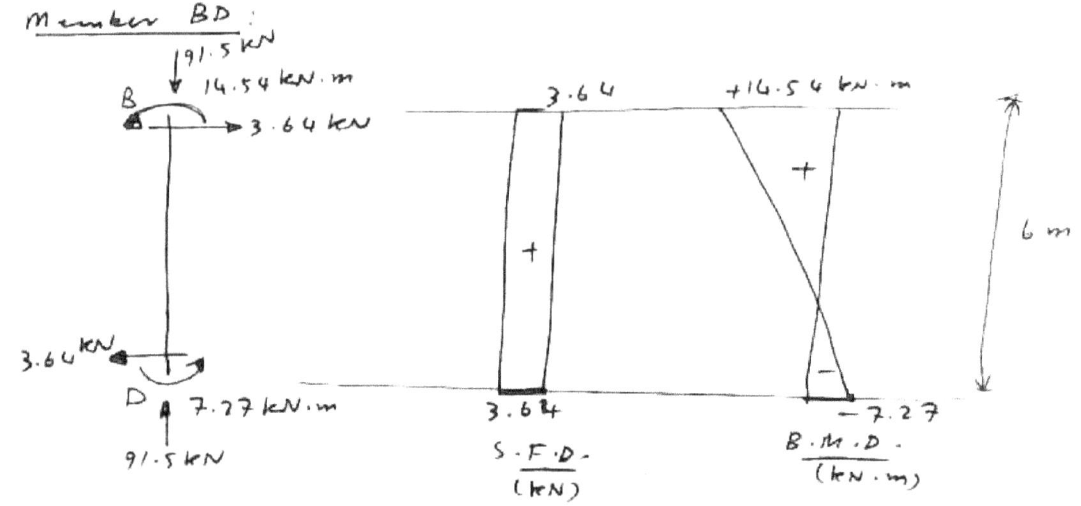

Member ABCE:

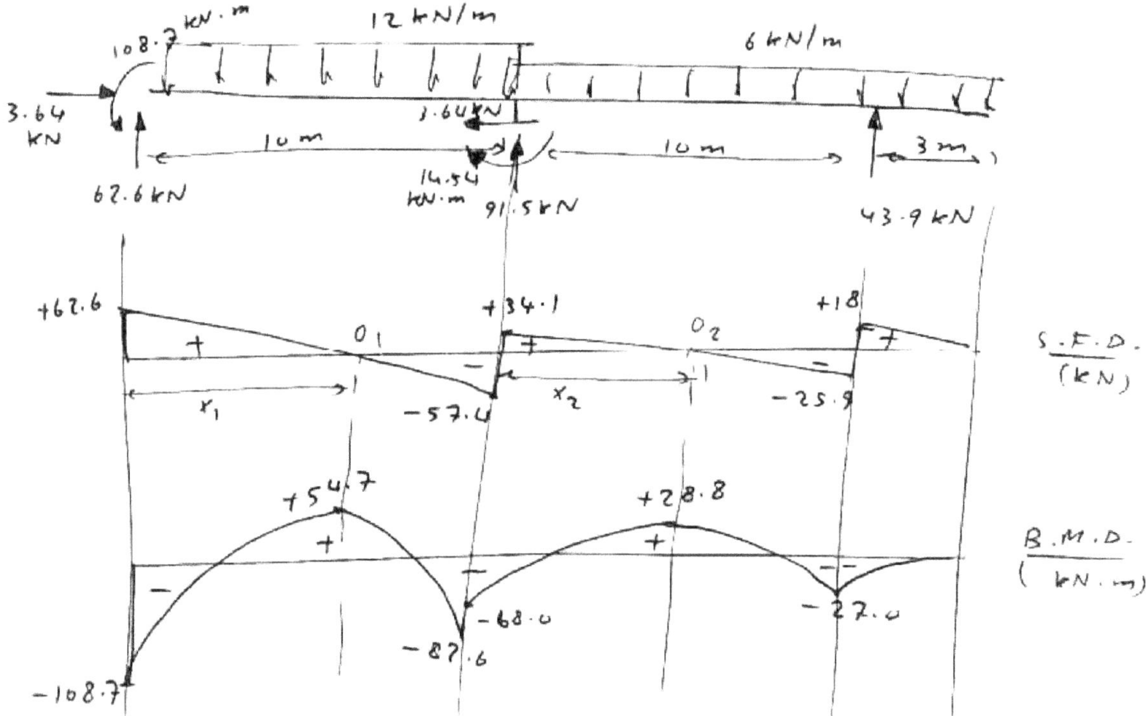

Locate the points of zero shear.

$$\frac{x_1}{62.6} = \frac{10 - x_1}{52.6}$$

$$52.4 x_1 = 626 - 62.6 x_1 \implies x_1 = 5.22 \text{ m}.$$

$$M_{o_1} = -108.7 + \frac{1}{2}(62.6)(5.22) = 54.7 \text{ kN m}.$$

$$\frac{x_2}{34.1} = \frac{10 - x_2}{25.9}$$

$$25.9 x_2 = 341 - 34.1 x_2 \implies x_2 = 5.68 \text{ m}.$$

$$M_{o_2} = -68.0 + \frac{1}{2}(34.1)(5.68) = 28.8 \text{ kN m}.$$

Example 13.4 :

Determine the member moments and reactions and draw the bending moment diagram for the frame shown in the figure.

Solution :

(see next page)

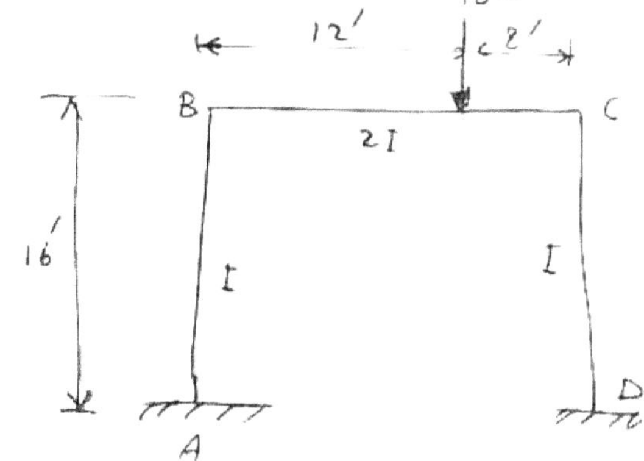

* Frames with Sidesway.

Example 13.4

Determine the member moments and reactions and draw the bending-moment diagram for the frame shown in the figure.

Solution:

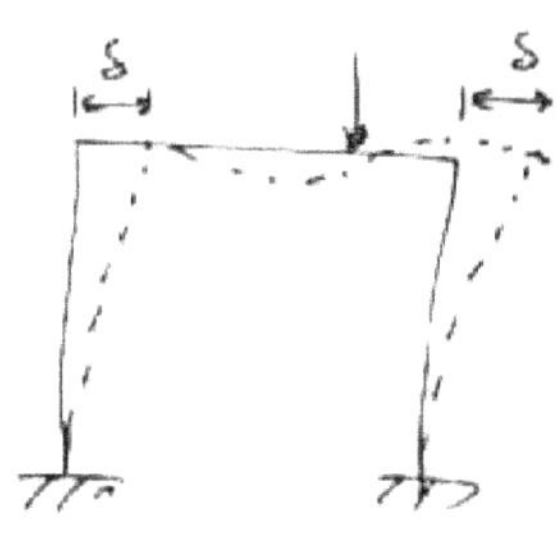

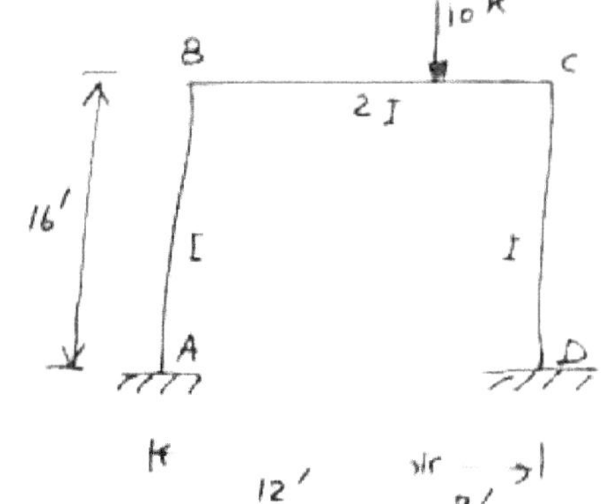

(1) Unknowns θ_B, θ_C, δ.

Note: We ignore axial deformation of flexural members, the ends B and C of member AB are assumed to move laterally the same amount,

i.e. $\delta_{BA} = \delta_{CD} = \delta$.

(2) Equations of Equilibrium:

Joint B : $M_{BA} + M_{BC} = 0$ ——— (1)

Joint C : $M_{CB} + M_{CD} = 0$ ——— (2)

At the base of the frame, take $\Sigma F_x = 0$.

$\xrightarrow{+} \Sigma F_x = 0$:

$$\boxed{H_A + H_D = 0}$$ (3)

"Shear Equation"

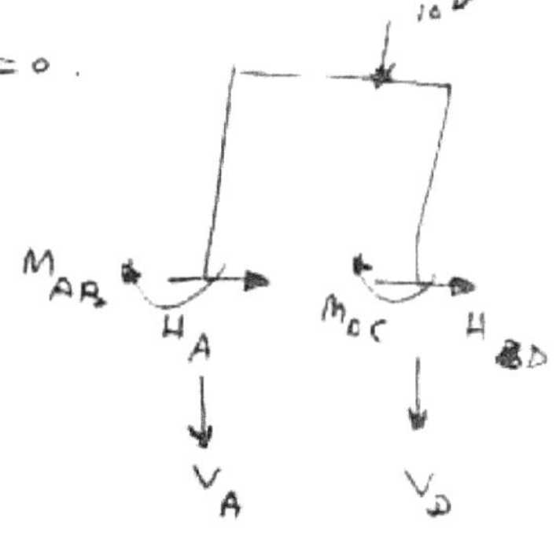

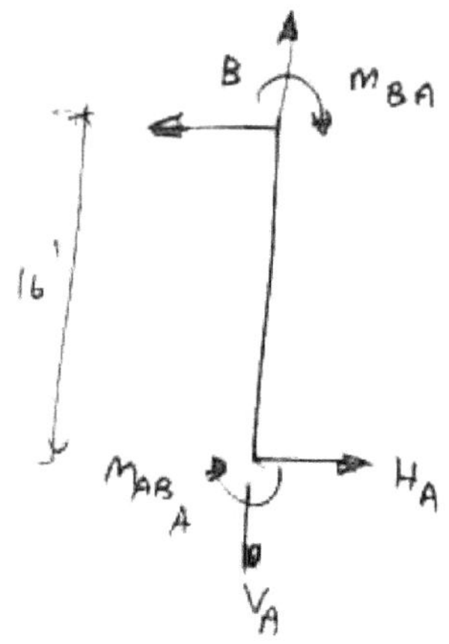

$+\circlearrowleft \Sigma M_B = 0$:

$$-M_{BA} - M_{AB} + H_A(16) = 0$$

$$\Rightarrow \quad H_A = \frac{M_{AB} + M_{BA}}{16}$$

$$+ \circlearrowleft \ \Sigma M_C = 0:$$

$$H_D (16) - m_{CD} - m_{DC} = 0$$

$$\implies \quad H_D = \frac{M_{CD} + M_{DC}}{16}$$

Substitute in the shear equation.

$$H_A + H_D = 0$$

$$\frac{m_{AB} + m_{BA}}{16} + \frac{m_{CD} + m_{DC}}{16} = 0$$

$$\implies \boxed{m_{AB} + m_{BA} + m_{CD} + m_{DC} = 0} \longrightarrow (3)$$

(3) Fixed-End Moments:

$$M^F_{BC} = -\frac{Pab^2}{L^2} = -10\frac{(12)(8)^2}{(20)^2} = -19.2 \ k\text{-}ft.$$

$$m^F_{CB} = +\frac{Pba^2}{L^2} = +10\frac{(8)(12)^2}{(20)^2} = +28.8 \ k\text{-}ft.$$

(4) Slope-Deflection Equations:

$$M_{AB} = \frac{2EI}{L}\left(2\theta_A + \theta_B - 3\frac{\delta_{AB}}{L}\right) \pm m^F_{AB}$$

$$= \frac{2EI}{16}\left(0 + \theta_B - 3\frac{\delta}{16}\right) \pm 0$$

$$\implies M_{AB} = 0.125\,EI\theta_B - 0.02344\,EI\delta$$

$$M_{BA} = \frac{2EI}{L}\left(2\theta_B + \theta_A - 3\frac{\delta_{AB}}{L}\right) \pm m^F_{BA}$$

$$= \frac{2EI}{16}\left(2\theta_B + 0 - 3\frac{\delta}{16}\right) \pm 0$$

$$\implies M_{BA} = 0.25\,EI\theta_B - 0.02344\,EI\delta$$

$$M_{BC} = \frac{2EI}{L}\left(2\theta_B + \theta_C - 3\frac{\delta_{BC}}{L}\right) \pm m^F_{BC}$$

$$= \frac{2E(2I)}{20}\left(2\theta_B + \theta_C - 0\right) - 19.2$$

$$\implies M_{BC} = 0.4\,EI\theta_B + 0.2\,EI\theta_C - 19.2$$

$$M_{CB} = \frac{2EI}{L}\left(2\theta_c + \theta_B - 3\frac{\delta_{CB}}{L}\right) \pm m^F_{CB}$$

$$= \frac{2E(2I)}{20}\left(2\theta_c + \theta_B - 0\right) + 28.8$$

$$\implies M_{CB} = 0.2\,EI\theta_B + 0.4\,EI\theta_c + 28.8$$

$$M_{CD} = \frac{2EI}{L}\left(2\theta_c + \theta_D - 3\frac{\delta_{CD}}{L}\right) \pm m^F_{CD}$$

$$= \frac{2EI}{16}\left(2\theta_c + 0 - \frac{3\delta}{16}\right) \pm 0$$

$$\implies M_{CD} = 0.25\,EI\theta_c - 0.02344\,EI\delta$$

$$M_{DC} = \frac{2EI}{L}\left(2\theta_D + \theta_c - 3\frac{\delta_{DC}}{L}\right) \pm m^F_{LD}$$

$$= \frac{2EI}{16}\left(0 + \theta_c - \frac{3\delta}{16}\right) \pm 0$$

$$\implies M_{DC} = 0.125\,EI\theta_c - 0.02344\,EI\delta$$

(5) <u>Substitute eqs. (4) into (2):</u>

$$M_{BA} + M_{BC} = 0$$

$$\implies \left(0.25\,EI\theta_B - 0.02344\,EI\delta\right) + \left(0.4\,EI\theta_B + 0.2\,EI\theta_c - 19.2\right) = 0$$

$$\implies \boxed{0.65\,EI\theta_B + 0.2\,EI\theta_c - 0.02344\,EI\delta = 19.2} \quad\text{———(1)}$$

$$M_{CB} + M_{CD} = 0$$

$$\implies \left(0.2\,EI\theta_B + 0.4\,EI\theta_c + 28.8\right) + \left(0.25\,EI\theta_c - 0.02344\,EI\delta\right) = 0$$

$$\implies \boxed{0.2\,EI\theta_B + 0.65\,EI\theta_c - 0.02344\,EI\delta = -28.8} \quad\text{———(2)}$$

$$M_{AB} + M_{BA} + M_{CD} + M_{DC} = 0$$

$$\left(0.125\,EI\theta_B - 0.02344\,EI\delta\right) + \left(0.25\,EI\theta_B - 0.02344\,EI\delta\right)$$

$$+ \left(0.25\,EI\theta_c - 0.02344\,EI\delta\right) + \left(0.125\,EI\theta_c - 0.02344\,EI\delta\right) = 0$$

$$\implies \boxed{0.375\,EI\theta_B + 0.375\,EI\theta_c - 0.09376\,EI\delta = 0} \quad\text{———(3)}$$

write the equations in matrix form:

$$EI \begin{bmatrix} 0.65 & 0.2 & -0.02344 \\ 0.2 & 0.65 & -0.02344 \\ 0.375 & 0.375 & -0.09376 \end{bmatrix} \begin{Bmatrix} \theta_B \\ \theta_C \\ \delta \end{Bmatrix} = \begin{Bmatrix} 19.2 \\ -28.8 \\ 0 \end{Bmatrix}$$

Use program SOLVE to get the results:

$$EI\theta_B = +46.088$$
$$EI\theta_C = -60.579$$
$$EI\delta = -57.956$$

(6) Substitute the results in step (4):

$$M_{AB} = 0.125\,(46.088) - 0.02344\,(-57.956)$$
$$= +7.12 \quad k\text{-ft.}$$

$$M_{BA} = 0.25\,(46.088) - 0.02344\,(-57.956)$$
$$= +12.88 \quad k\text{-ft.}$$

$$M_{BC} = 0.4\,(46.088) + 0.2\,(-60.579) - 19.2$$
$$= -12.88 \quad k\text{-ft.}$$

$$M_{CB} = 0.2\,(46.088) + 0.4\,(-60.579) + 28.8$$
$$= +13.79 \quad k\text{-ft.}$$

$$M_{CD} = 0.25\,(-60.579) - 0.02344\,(-57.956)$$
$$= -13.79 \quad k\text{-ft.}$$

$$M_{DC} = 0.125\,(-60.579) - 0.02344\,(-57.956)$$
$$= -6.21 \quad k\text{-ft.}$$

(7) Use Statics to find the shears and reactions:

$$H_A = \frac{M_{AB} + M_{BA}}{16} = \frac{7.12 + 12.88}{16} = 1.25 \ k.$$

$$H_D = \frac{M_{CD} + M_{DC}}{16} = \frac{-13.79 - 6.21}{16} = -1.25 \ k.$$

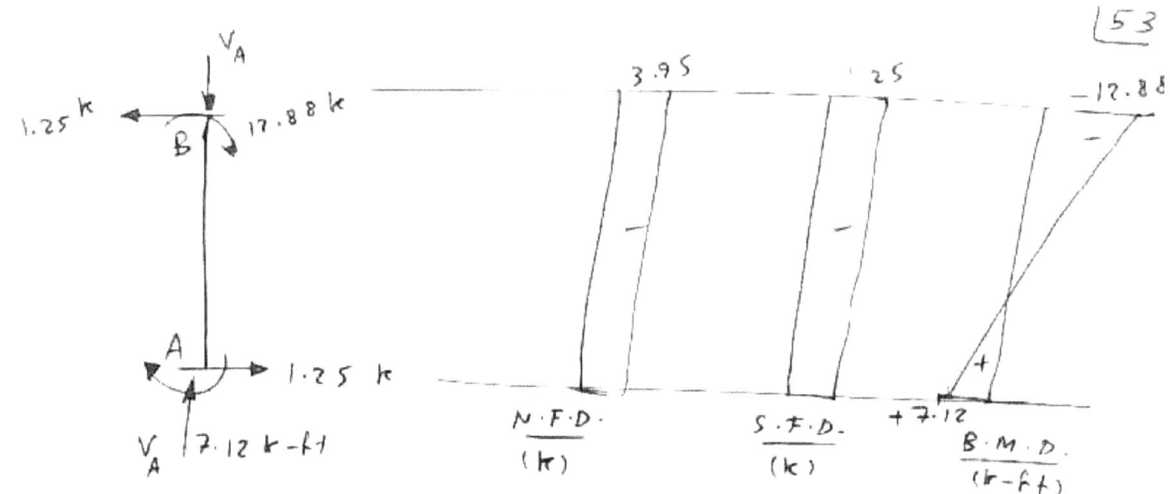

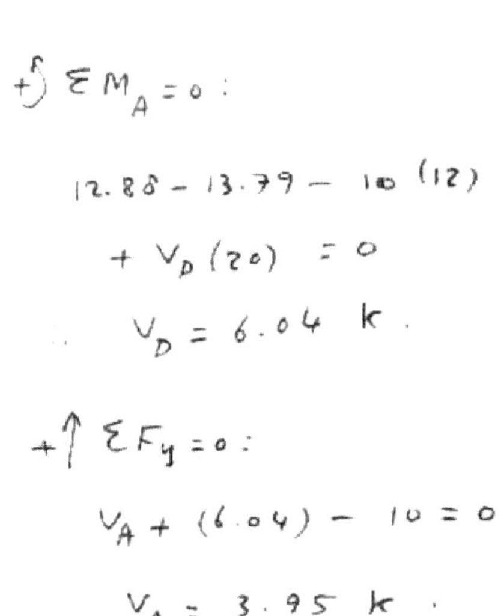

$$+ \circlearrowleft \ \Sigma M_A = 0:$$

$$12.88 - 13.79 - 10\,(12)$$

$$+ V_D\,(20) = 0$$

$$V_D = 6.04 \ k$$

$$+\uparrow \ \Sigma F_y = 0:$$

$$V_A + (6.04) - 10 = 0$$

$$V_A = 3.95 \ k$$

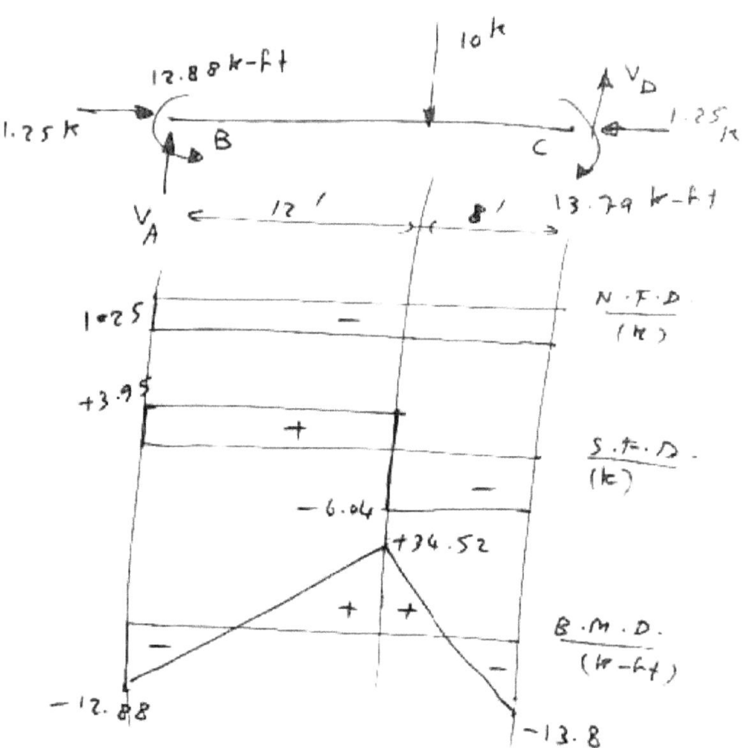

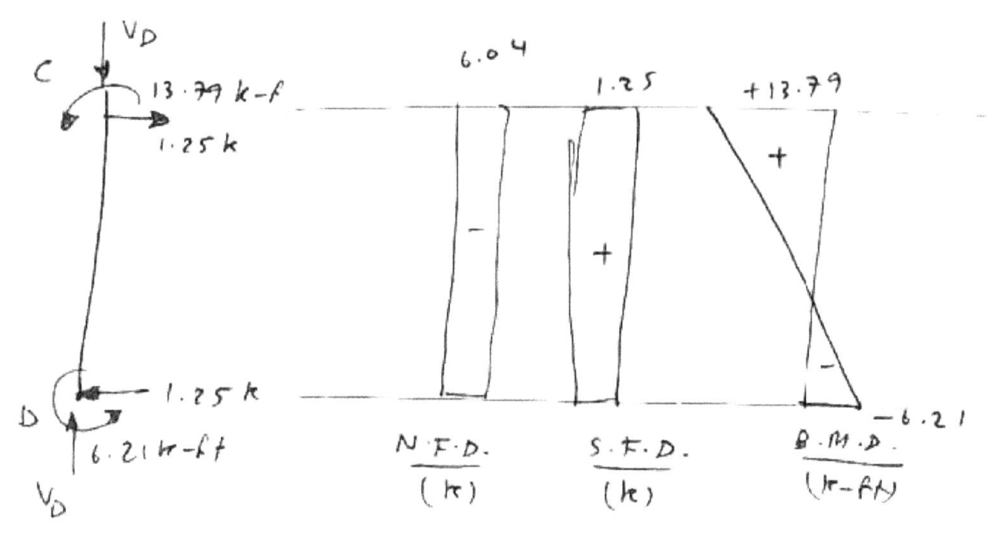

Example 13.5:

Determine the member moments and the reactions and draw the bending moment diagram for the multi-storey frame shown in the figure.

EI = constant.

Solution:

(1) Unknowns:

$$\theta_B, \theta_C, \theta_D, \theta_E, \delta_1, \delta_2$$

Note: There are two joint translations; one at each level.

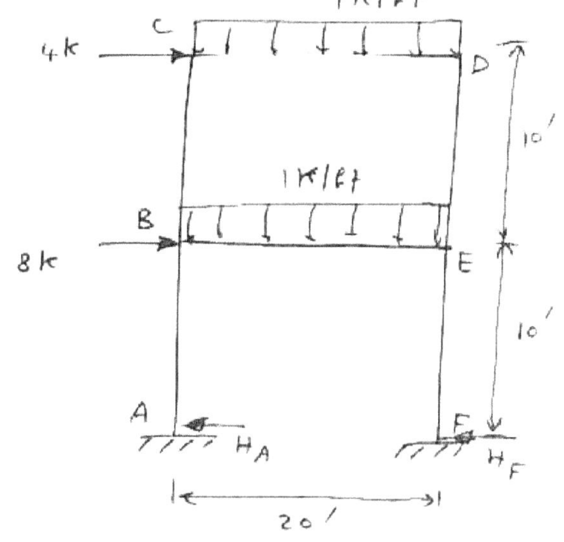

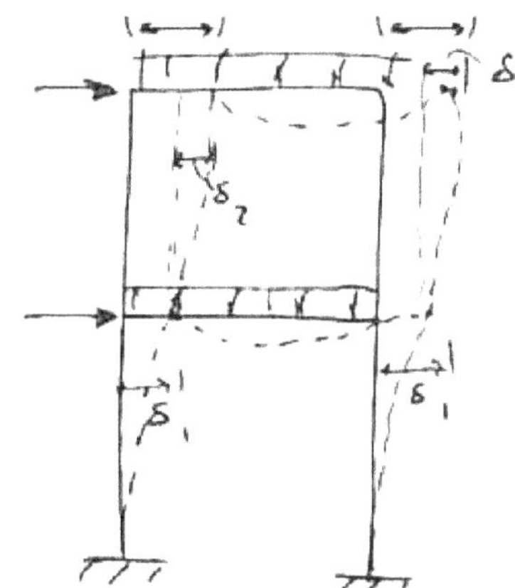

δ_1, δ_2 : displacements of one end of the other relative to the other end

* Neglect axial deformation in flexural members.

(2) Equations of Equilibrium:

Joint B. $M_{BA} + M_{BE} + M_{BC} = 0$ ——————— (1)

Joint C. $M_{CB} + M_{CD} = 0$ ——————— (2)

Joint D: $M_{DC} + M_{DE} = 0$ ——————— (3)

Joint E: $M_{EF} + M_{EB} + M_{ED} = 0$ ——————— (4)

We need two shear equations. (one at each level).

At the base of the frame:

$\xrightarrow{+} \Sigma F_x = 0.$ $4 + 8 - H_A - H_F = 0$

$$\implies \boxed{H_A + H_F = 12}$$

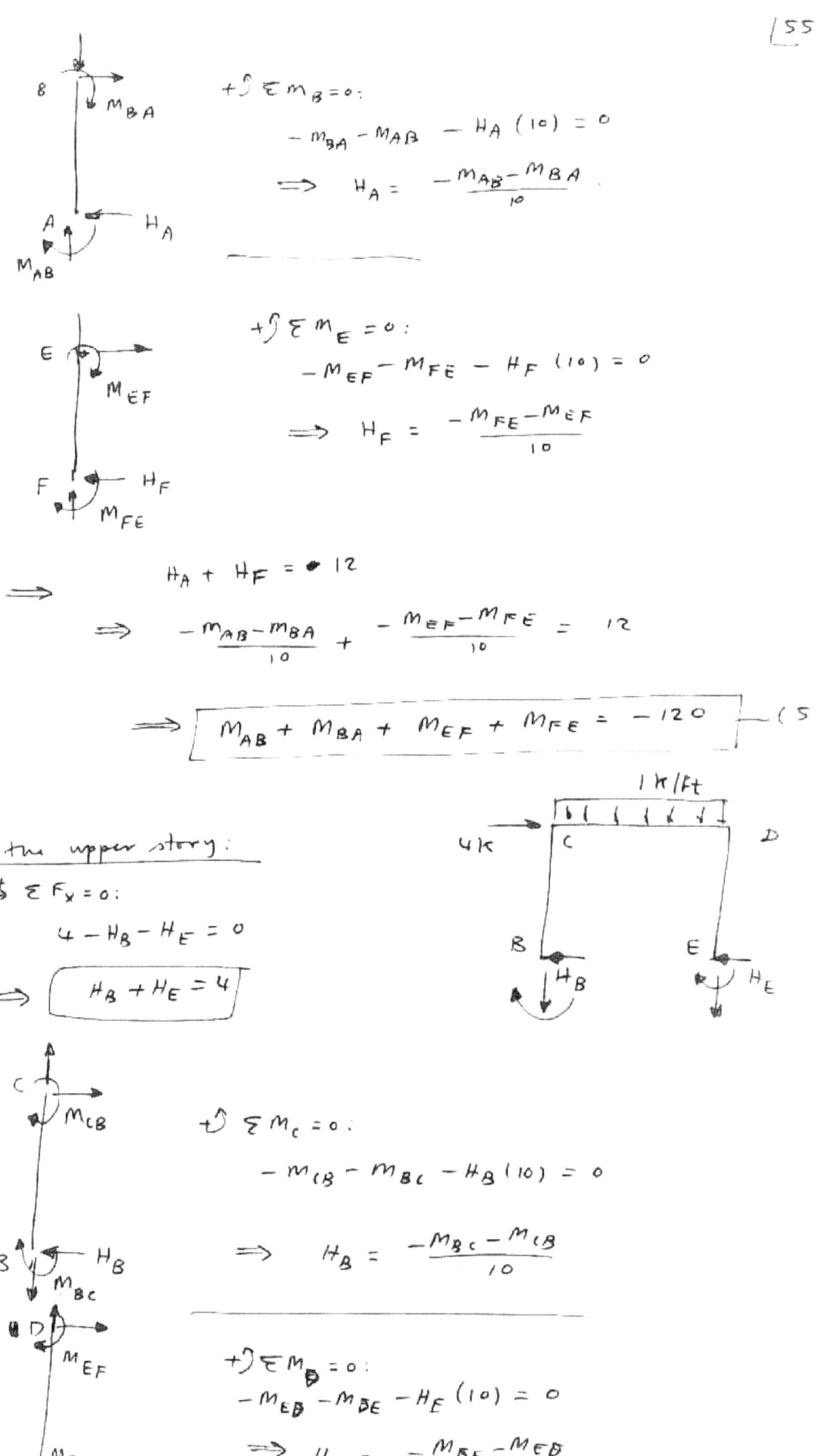

$+\circlearrowleft \Sigma m_B = 0:$

$$-m_{BA} - M_{AB} - H_A(10) = 0$$

$$\implies H_A = \frac{-M_{AB} - M_{BA}}{10}$$

$+\circlearrowleft \Sigma m_E = 0:$

$$-M_{EF} - M_{FE} - H_F(10) = 0$$

$$\implies H_F = \frac{-M_{FE} - M_{EF}}{10}$$

$$\implies H_A + H_F = 12$$

$$\implies \frac{-m_{AB} - m_{BA}}{10} + \frac{-M_{EF} - M_{FE}}{10} = 12$$

$$\implies \boxed{M_{AB} + M_{BA} + M_{EF} + M_{FE} = -120} \quad (5)$$

At the upper story:

$\xrightarrow{} \Sigma F_x = 0:$

$$4 - H_B - H_E = 0$$

$$\implies \boxed{H_B + H_E = 4}$$

$+\circlearrowleft \Sigma M_c = 0:$

$$-m_{CB} - m_{BC} - H_B(10) = 0$$

$$\implies H_B = \frac{-M_{BC} - M_{CB}}{10}$$

$+\circlearrowleft \Sigma M_D = 0:$

$$-m_{ED} - m_{DE} - H_E(10) = 0$$

$$\implies H_E = \frac{-M_{DE} - M_{ED}}{10}$$

$$H_B + H_E = 0$$

$$\implies \quad -\frac{M_{BC} - M_{CB}}{10} + -\frac{M_{DE} - M_{ED}}{10} = 4$$

$$\implies \boxed{M_{BC} + M_{CB} + M_{DE} + M_{ED} = -40} \quad\text{——— (6)}$$

(3) $\underline{\text{Slope-Deflection Equations}}$:

$$M_{AB} = 2\frac{EI}{L}\left(2\theta_A + \theta_B - 3\frac{\delta_{AB}}{L}\right) \pm M_{AB}^F$$

$$= 2\frac{EI}{10}\left(0 + \theta_B - 3\frac{\delta_1}{10}\right) \pm 0$$

$$\implies \boxed{M_{AB} = 0.2\,EI\,\theta_B - 0.06\,EI\,\delta_1}$$

$$M_{BA} = 2\frac{EI}{L}\left(2\theta_B + \theta_A - 3\frac{\delta_{AB}}{L}\right) \pm M_{BA}^F$$

$$= 2\frac{EI}{10}\left(2\theta_B + 0 - 3\frac{\delta_1}{10}\right) \pm 0$$

$$\implies \boxed{M_{BA} = 0.4\,EI\,\theta_B - 0.06\,EI\,\delta_1}$$

$$M_{BE} = 2\frac{EI}{L}\left(2\theta_B + \theta_E - 3\frac{\delta_{BE}}{L}\right) \pm \frac{wL^2}{12}$$

$$= 2\frac{EI}{20}\left(2\theta_B + \theta_E - 0\right) - \frac{(1)(20)^2}{12}$$

$$\implies \boxed{M_{BE} = 0.2\,EI\,\theta_B + 0.1\,EI\,\theta_E - 33.333}$$

$$M_{EB} = 2\frac{EI}{L}\left(2\theta_E + \theta_B - 3\frac{\delta_{EB}}{L}\right) \pm \frac{wL^2}{12}$$

$$= 2\frac{EI}{20}\left(2\theta_E + \theta_B - 0\right) + \frac{(1)(20)^2}{12}$$

$$\implies \boxed{M_{EB} = 0.2\,EI\,\theta_E + 0.1\,EI\,\theta_B + 33.333}$$

$$M_{BC} = 2\frac{EI}{L}\left(2\theta_B + \theta_C - 3\frac{\delta_{BC}}{L}\right) \pm M_{BC}^F$$

$$= 2\frac{EI}{10}\left(2\theta_B + \theta_C - 3\frac{\delta_2}{10}\right) \pm 0$$

$$\implies \boxed{M_{BC} = 0.4\,EI\,\theta_B + 0.2\,EI\,\theta_C - 0.06\,EI\,\delta_2}$$

$$M_{CB} = 2\frac{EI}{10}\left(2\theta_C + \theta_B - 3\frac{\delta_2}{10}\right) \pm 0$$

$$\implies \boxed{M_{CB} = 0.4\,EI\,\theta_C + 0.2\,EI\,\theta_B - 0.06\,EI\,\delta_2}$$

$$M_{CD} = \frac{2EI}{20}\left(2\theta_C + \theta_D - 0\right) - \frac{(1)(20)^2}{12}$$

$$\Rightarrow \boxed{M_{CD} = 0.2\,EI\,\theta_C + 0.1\,EI\,\theta_D - 33.333}$$

$$\boxed{M_{DC} = 0.2\,EI\,\theta_D + 0.1\,EI\,\theta_C + 33.333}$$

$$M_{DE} = \frac{2EI}{L}\left(2\theta_D + \theta_E - 3\frac{\delta_{DE}}{L}\right) \pm M^F_{DE}$$

$$= \frac{2EI}{10}\left(2\theta_D + \theta_E - 3\frac{\delta_2}{10}\right) \pm 0$$

$$\Rightarrow \boxed{M_{DE} = 0.4\,EI\,\theta_D + 0.2\,EI\,\theta_E - 0.06\,EI\,\delta_2}$$

$$\boxed{M_{ED} = 0.4\,EI\,\theta_E + 0.2\,EI\,\theta_D - 0.06\,EI\,\delta_2}$$

$$M_{EF} = \frac{2EI}{L}\left(2\theta_E + \theta_F - 3\frac{\delta_{EF}}{L}\right) \pm M^F_{EF}$$

$$= \frac{2EI}{10}\left(2\theta_E + 0 - 3\frac{\delta_1}{10}\right) \pm 0$$

$$\Rightarrow \boxed{M_{EF} = 0.4\,EI\,\theta_E - 0.06\,EI\,\delta_1}$$

$$\boxed{M_{FE} = 0.2\,EI\,\theta_E - 0.06\,EI\,\delta_1}$$

---

(4) Substitute eqs. (3) into step (2):

$$M_{BA} + M_{BE} + M_{BC} = 0$$

$$\Rightarrow 0.4\,EI\,\theta_B - 0.06\,EI\,\delta_1 + 0.2\,EI\,\theta_B + 0.1\,EI\,\theta_E - 33.333$$
$$+ 0.4\,EI\,\theta_B + 0.2\,EI\,\theta_C - 0.06\,EI\,\delta_2 = 0$$

$$\Rightarrow \boxed{\begin{aligned}EI\,\theta_B + 0.2\,EI\,\theta_C + 0.1\,EI\,\theta_E - 0.06\,EI\,\delta_1 \\ - 0.06\,EI\,\delta_2 = 33.333\end{aligned}} \quad ①$$

$$M_{CB} + M_{CD} = 0$$

$$\Rightarrow 0.4\,EI\,\theta_C + 0.2\,EI\,\theta_B - 0.06\,EI\,\delta_2 + 0.2\,EI\,\theta_C + 0.1\,EI\,\theta_D$$
$$- 33.333 = 0$$

$$\Rightarrow \boxed{0.2\,EI\,\theta_B + 0.6\,EI\,\theta_C + 0.1\,EI\,\theta_D - 0.06\,EI\,\delta_2 = 33.333} \quad ②$$

$$M_{DC} + M_{DE} = 0$$

$$\Rightarrow \quad 0.2EI\theta_D + 0.1EI\delta_c + 33.333 + 0.4EI\theta_D + 0.2EI\theta_E - 0.06EI\delta_2 = 0$$

$$\Rightarrow \quad \boxed{0.1EI\theta_C + 0.6EI\theta_D + 0.2EI\theta_E - 0.06EI\delta_2 = -33.333} \quad —③$$

$$M_{EF} + M_{EB} + M_{ED} = 0$$

$$\Rightarrow \quad 0.4EI\theta_E - 0.06EI\delta_1 + 0.2EI\theta_E + 0.1EI\theta_B + 33.333$$
$$\qquad + 0.4EI\theta_E + 0.2EI\theta_D - 0.06EI\delta_2 = 0$$

$$\Rightarrow \quad \boxed{0.1EI\theta_B + 0.2EI\theta_D + EI\theta_E - 0.06EI\delta_1 - 0.06EI\delta_2 = -33.333} \quad —④$$

$$M_{AB} + M_{BA} + M_{EF} + M_{FE} = -120$$

$$\Rightarrow \quad 0.2EI\theta_B - 0.06EI\delta_1 + 0.4EI\theta_B - 0.06EI\delta_1$$
$$\qquad + 0.4EI\theta_E - 0.06EI\delta_1 + 0.2EI\theta_E - 0.06EI\delta_1 = -120$$

$$\Rightarrow \quad \boxed{0.6EI\theta_B + 0.6EI\theta_E - 0.24EI\delta_1 = -120} \quad —⑤$$

$$M_{BC} + M_{CB} + M_{DE} + M_{ED} = -40$$

$$0.4EI\theta_B + 0.2EI\theta_C - 0.06EI\delta_2 + 0.4EI\theta_C + 0.2EI\theta_B - 0.06EI\delta_2$$
$$\qquad + 0.4EI\theta_D + 0.2EI\theta_E - 0.06EI\delta_2 + 0.4EI\theta_E + 0.2EI\theta_D - 0.06EI\delta_2$$
$$\qquad\qquad = -40$$

$$\Rightarrow \quad \boxed{0.6EI\theta_B + 0.6EI\theta_C + 0.6EI\theta_D + 0.6EI\theta_E - 0.24EI\delta_2 = -40} \quad —⑥$$

* see next page for continuation !

Write the equations in <u>matrix form</u>:

$$\frac{1}{EI}\begin{bmatrix} 1.0 & 0.2 & 0 & 0.1 & -0.06 & -0.06 \\ 0.2 & 0.6 & 0.1 & 0 & 0 & -0.06 \\ 0 & 0.1 & 0.6 & 0.2 & 0 & -0.06 \\ 0.1 & 0 & 0.2 & 1.0 & -0.06 & -0.06 \\ 0.6 & 0 & 0 & 0.6 & -0.24 & 0 \\ 0.6 & 0.6 & 0.6 & 0.6 & 0 & -0.24 \end{bmatrix} \begin{Bmatrix} \theta_B \\ \theta_C \\ \theta_D \\ \theta_E \\ \delta_1 \\ \delta_2 \end{Bmatrix} = \begin{Bmatrix} 33.333 \\ 33.333 \\ -33.333 \\ -33.333 \\ -120.0 \\ -40.0 \end{Bmatrix}$$

Use program <u>SOLVE</u> to get:

$$EI\theta_B = 113.86$$
$$EI\theta_C = 104.28$$
$$EI\theta_D = -9.54$$
$$EI\theta_E = 65.08$$
$$EI\delta_1 = 947.37$$
$$EI\delta_2 = 850.88$$

---

(5) Substitute the results into step (3):

$M_{AB} = 0.2(113.86) - 0.06(947.37) = -34.07$ k-ft.

$M_{BA} = 0.4(113.86) - 0.06(947.37) = -11.30$ k-ft.

$M_{BE} = 0.2(113.86) + 0.1(65.08) - 33.333 = -4.05$ k-ft.

$M_{EB} = 0.2(65.08) + 0.1(113.86) + 33.333 = 57.74$ k-ft.

$M_{BC} = 0.4(113.86) + 0.2(104.28) - 0.06(850.88) = 15.35$ k-ft

$M_{CB} = 0.4(104.28) + 0.2(113.86) - 0.06(850.88) = 13.43$ k-ft

$M_{CD} = 0.2(104.28) + 0.1(-9.54) - 33.333 = -13.43$ k-ft

$M_{DC} = 0.2(-9.54) + 0.1(104.28) + 33.333 = 41.85$ k-ft.

$M_{DE} = 0.4(-9.54) + 0.2(65.08) - 0.06(850.88) = -41.25$ k-ft.

$M_{ED} = 0.4(65.08) + 0.2(-9.54) - 0.06(850.88) = -26.93$ k-ft.

$$M_{EF} = 0.4(65.08) - 0.06(947.37) = -30.21 \text{ k-ft}$$

$$M_{FE} = 0.2(65.08) - 0.06(947.37) = -43.23 \text{ k-ft}$$

(6) Use <u>Statics</u> to get the shears and reactions:

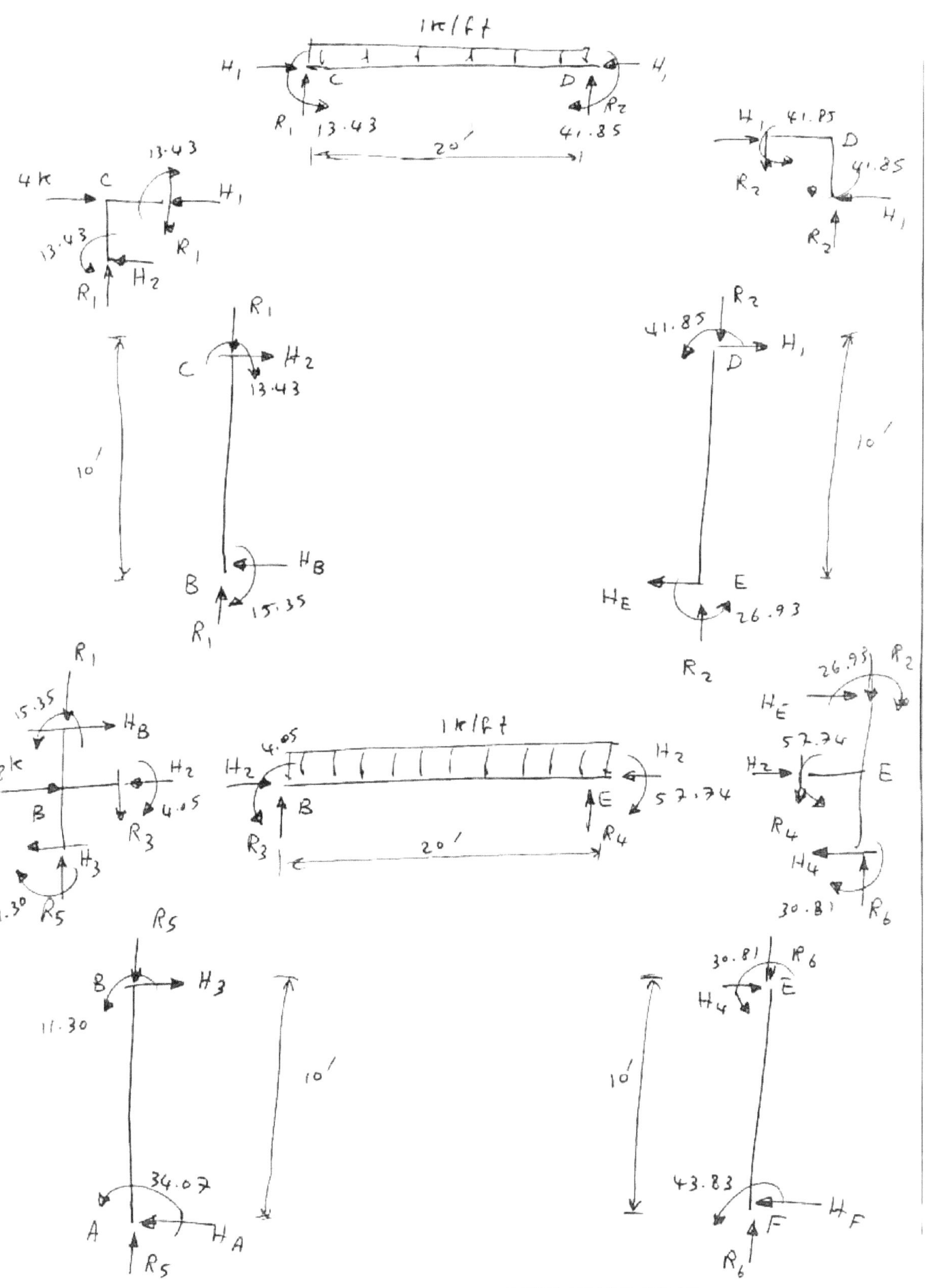

$$H_A = \frac{-M_{AB} \quad M_{BA}}{10} = \frac{+34.07 + 11.30}{10} = 4.54 \text{ k.}$$

$$H_F = \frac{-M_{FE} - M_{EF}}{10} = \frac{+30.81 + 43.83}{10} = 7.46 \text{ k.}$$

$$H_B = \frac{-M_{BC} - M_{CB}}{10} = \frac{-15.35 - 13.43}{10} = -2.88 \text{ k.}$$

$$H_E = \frac{-M_{DE} - M_{ED}}{10} = \frac{+41.85 + 26.73}{10} = 6.88 \text{ k.}$$

Member CD : $+\circlearrowright \Sigma M_C = 0$.

$$-41.85 + 13.43 + R_2(20) - 1(20)(10) = 0$$

$$\Rightarrow R_2 = 11.42 \text{ k.}$$

$+\uparrow \Sigma F_y = 0.$ $R_1 + 11.42 - 1(20) = 0$

$$\Rightarrow R_1 = 8.58 \text{ k.}$$

Member BE : $+\circlearrowright \Sigma M_B = 0$.

$$-57.74 + 4.05 + R_4(20) - 1(20)(10) = 0$$

$$\Rightarrow R_4 = 12.68 \text{ k.}$$

$+\uparrow \Sigma F_y = 0$: $R_3 + 12.68 - 1(20) = 0$

$$\Rightarrow R_3 = 7.32 \text{ k.}$$

Joint B . $+\uparrow \Sigma F_y = 0$: $R_5 - 7.32 - 8.58 = 0$

$$R_5 = 15.9 \text{ k.}$$

Joint E : $+\uparrow \Sigma F_y = 0$. $R_6 - 12.68 - 11.42 = 0$

$$\Rightarrow R_6 = 24.1 \text{ k.}$$

Member AB : $H_3 = H_A = 4.54 \text{ k.}$

Member EF . $H_4 = H_F = 7.46 \text{ k.}$

Joint B : $\xrightarrow{} \Sigma F_x = 0$: $-4.54 - H_2 + (-2.88) + 8 = 0$

$$\Rightarrow H_2 = +0.58 \text{ k} \approx \underline{0}.$$

Joint E . $\xrightarrow{} \Sigma F_x = 0$: $-7.46 + H_2 + 6.88 = 0$

$$\Rightarrow H_2 = +0.58 \text{ k} \approx \underline{0}$$
$$(\text{Check } \checkmark)$$

(Axial deformation can be neglected.)

Member BC . $H_2 = H_B = -2.88$ k

Member DE . $H_1 = H_E = 6.88$ k.

Joint C . $\xrightarrow{+} \Sigma F_x = 0$. $-(-2.88) - H_1 + 4 = 0$

$$H_1 = 6.88 \text{ k .}$$

Joint D . $H_1 = 6.88$ (check ✓).

(7) Draw $S \cdot F \cdot D$. and $B \cdot M \cdot D$.

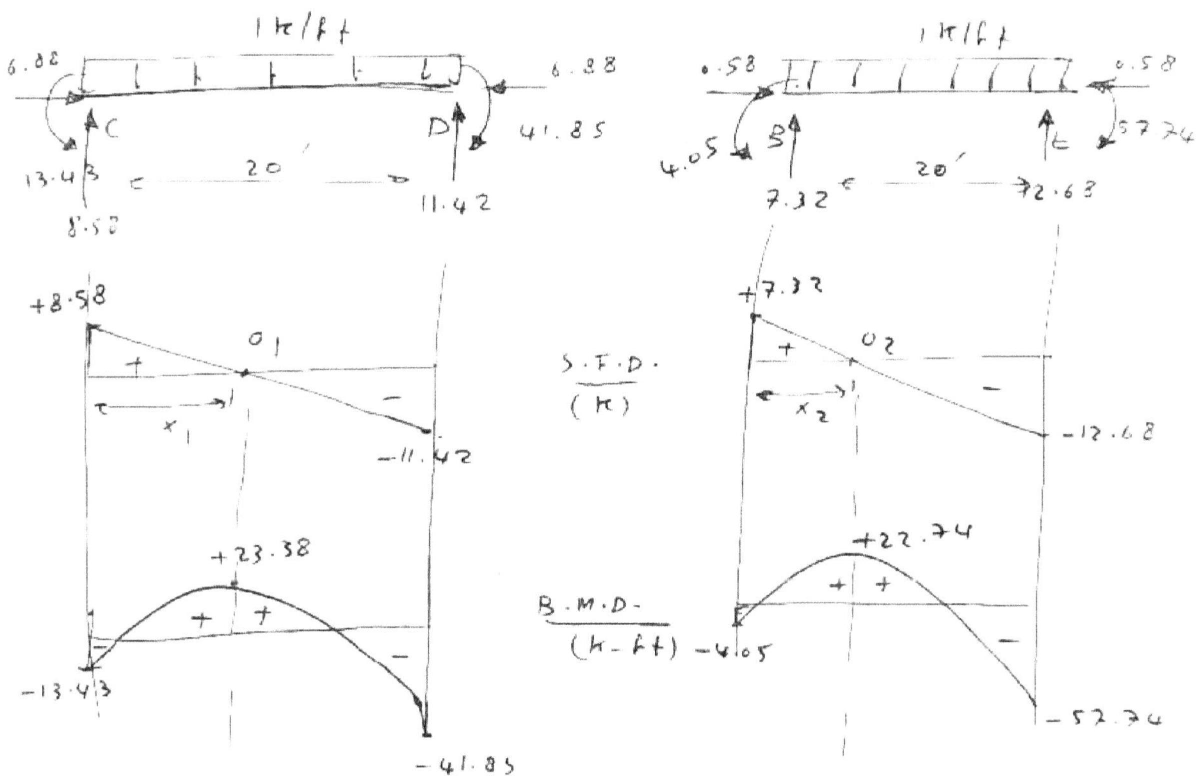

Locate points O_1 and O_2 of zero shear.

$$\frac{x_1}{8.58} = \frac{20 - x_1}{11.42} \implies 11.42 x_1 = 20(8.58) - 8.58 x_1$$
$$\implies x_1 = 8.58 \text{ ft .}$$

$$\frac{x_2}{7.32} = \frac{20 - x_2}{12.68} \implies 12.68 x_2 = 20(7.32) - 7.32 x_2$$
$$\implies x_2 = 7.32 \text{ ft .}$$

$$M_{O_1} = -13.43 + \tfrac{1}{2}(8.58)(8.58) = 23.38 \text{ k-ft .}$$

$$M_{O_2} = -4.05 + \tfrac{1}{2}(7.32)(7.32) = 22.74 \text{ k-ft .}$$

B.M.D only is shown for other members:

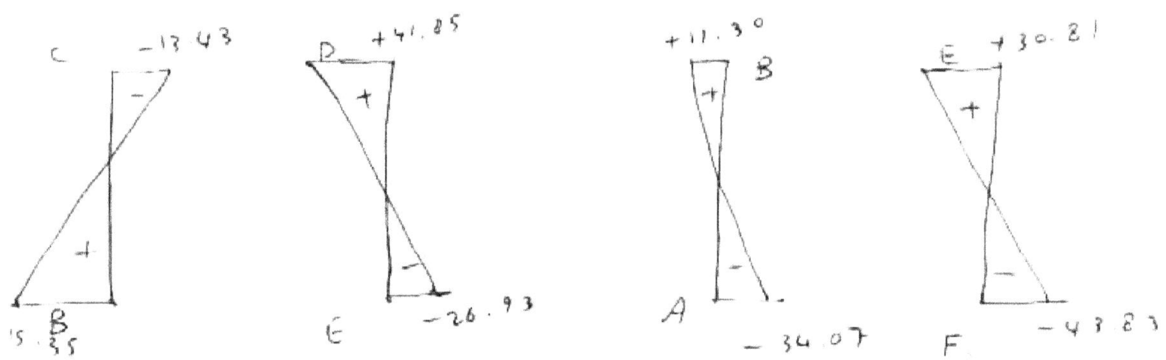

* Final B.M.D is shown as follows. (k-ft).

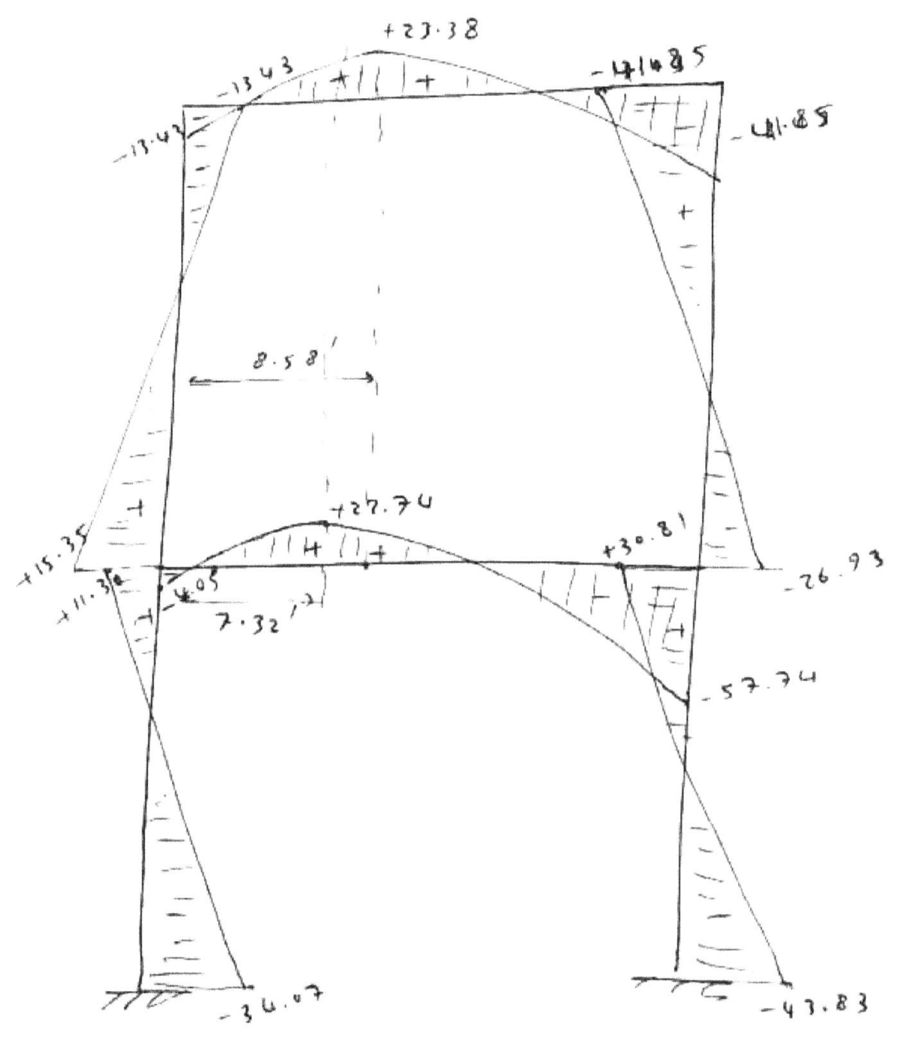

Frames with Sloping Members:

Example 1:

Draw the bending moment diagram for the frame shown in the figure.

Solution:

No. of degrees of freedom = 3

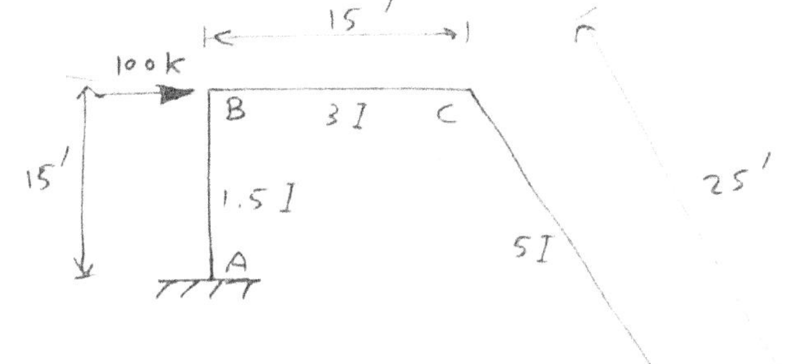

(1) Unknowns:

θ_B, θ_C, δ

$\delta_{AB} = \delta$

(clockwise member rotation).

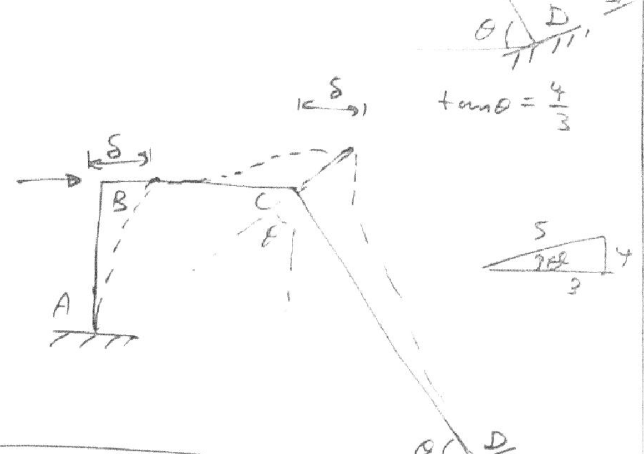

$\tan\theta = \dfrac{\delta}{\delta_{BC}}$

$$\Rightarrow \delta_{BC} = \frac{-\delta}{\tan\theta} = -\frac{3}{4}\delta = -0.75\delta. \quad \text{(counterclockwise member rotation)}.$$

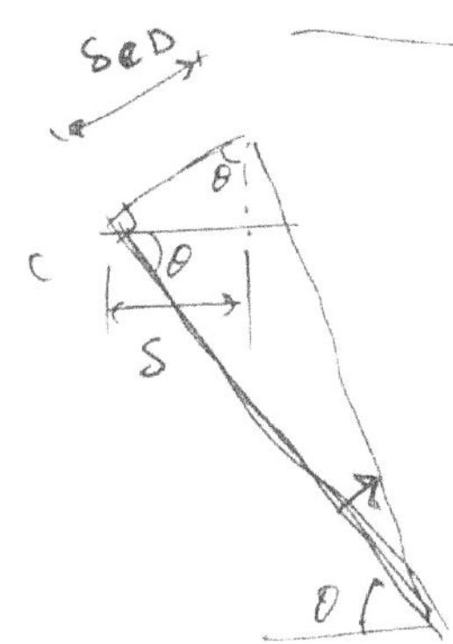

$\sin\theta = \dfrac{\delta}{\delta_{CD}}$

$$\Rightarrow \delta_{CD} = +\frac{\delta}{\sin\theta} = \frac{5}{4}\delta = +1.25\,\delta. \quad \text{(clockwise member rotation)}.$$

(2) Write equilibrium equations at the joints which are free to rotate.

At joint B: $\boxed{M_{BA} + M_{BC} = 0}$ —————— (1)

At joint C: $\boxed{M_{CB} + M_{CD} = 0}$ —————————— (2)

Shear Equation:

$+\circlearrowleft \; \Sigma M_O = 0$: (Take Moment about O)!!

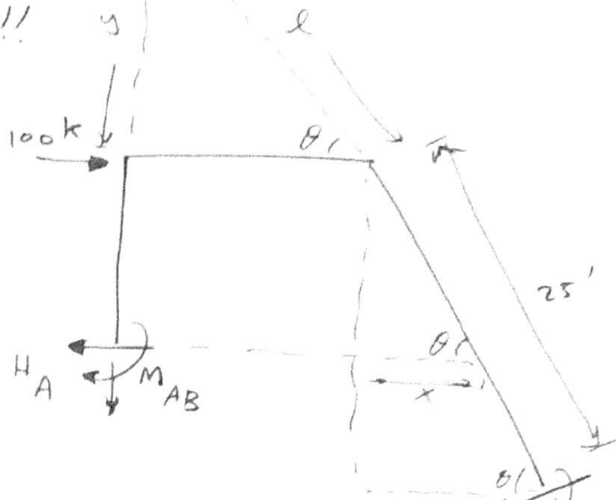

$\tan\theta = \dfrac{15}{x} = \dfrac{4}{3}$

$x = 11.25'$.

$\tan\theta = \dfrac{y}{15} = \dfrac{4}{3}$

$y = 20'$.

$\ell = \sqrt{(20)^2 + (15)^2} = 25'$.

$+\circlearrowleft \; \Sigma m_O = 0$:

$-M_{AB} - M_{DC} + 100(20) - H_A(35) - H_D(50) = 0$ ——————————— (3)

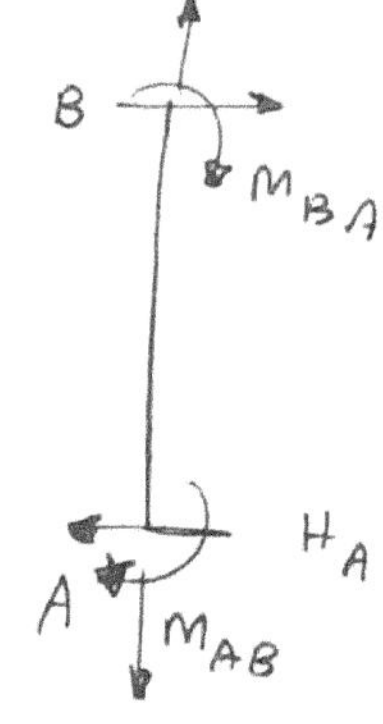

$+\circlearrowleft \; \Sigma m_B = 0$:

$-M_{AB} - M_{BA} - H_A(15) = 0$

$\Rightarrow \qquad H_A = \dfrac{-M_{AB} - M_{BA}}{15}$

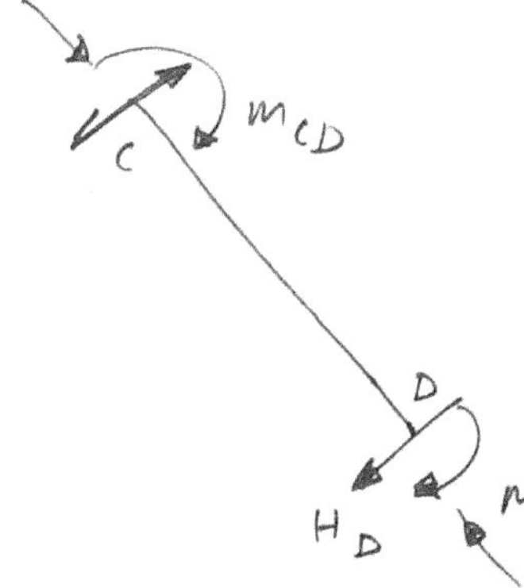

$+\circlearrowleft \; \Sigma m_C = 0$:

$-M_{CD} + M_{DC} + H_D(25) = 0$

$H_D = \dfrac{-M_{CD} - M_{DC}}{25}$.

$\Rightarrow -M_{AB} - M_{DC} + 2000 - 35\left(\dfrac{-M_{AB} - M_{BA}}{15}\right) - 50\left(\dfrac{-M_{CD} - M_{DC}}{25}\right) = 0$

$M_{AB} + M_{DC} - 2000 - 2.333\,M_{AB} - 2.333\,M_{BA} - 2\,M_{CD} - 2\,M_{DC} = 0$

$$-1.333\, m_{AB} - 2.333\, m_{BA} - m_{BC} - 2\, m_{CD} = 2000 \qquad (3)$$

(3) Slope – Deflection Equations:

$$m_{AB} = \frac{2EI}{L}\left(2\theta_A + \theta_B - 3\frac{S_{AB}}{L}\right) \pm M_{AB}^{F}$$

$$= \frac{2E(1.5I)}{15}\left(0 + \theta_B - \frac{3\delta}{15}\right) \pm 0$$

$$\Rightarrow \quad \boxed{M_{AB} = 0.2\, EI\,\theta_B - 0.04\, EI\, S}$$

$$M_{BA} = \frac{2EI}{L}\left(2\theta_B + \theta_A - 3\frac{S_{AB}}{L}\right) \pm M_{BA}^{F}$$

$$= \frac{2E(1.5I)}{15}\left(2\theta_B + 0 - \frac{3\delta}{15}\right) \pm 0$$

$$\Rightarrow \quad \boxed{M_{BA} = 0.4\, EI\,\theta_B - 0.04\, EI\, S}$$

$$M_{BC} = \frac{2EI}{L}\left(2\theta_B + \theta_C - 3\frac{S_{BC}}{L}\right) \pm M_{BC}^{F}$$

$$= \frac{2E(3I)}{15}\left(2\theta_B + \theta_C - \frac{3(-0.75S)}{15}\right) \pm 0$$

$$\Rightarrow \quad \boxed{M_{BC} = 0.8\, EI\,\theta_B + 0.4\, EI\,\theta_C + 0.06\, EI\, S}$$

$$M_{CB} = \frac{2EI}{L}\left(2\theta_C + \theta_B - 3\frac{S_{CB}}{L}\right) \pm M_{CB}^{F}$$

$$= \frac{2E(3I)}{15}\left(2\theta_C + \theta_B - \frac{3(-0.75S)}{15}\right) \pm M_{CB}^{F}$$

$$\Rightarrow \quad \boxed{M_{CB} = 0.8\, EI\,\theta_C + 0.4\, EI\,\theta_B + 0.06\, EI\, S}$$

$$M_{CD} = \frac{2E(5I)}{25}\left(2\theta_C + \theta_D - 3\frac{S_{CD}}{L}\right) \pm M_{CD}^{F}$$

$$= \frac{10EI}{25}\left(2\theta_C + 0 - \frac{3(1.25S)}{25}\right) \pm 0$$

$$\Rightarrow \quad \boxed{M_{CD} = 0.8\, EI\,\theta_C - 0.06\, EI\, S}$$

$$M_{DC} = \frac{2EI}{L}\left(2\theta_D + \theta_C - \frac{3\,S_{DC}}{L}\right) \pm M_{DC}^F$$

$$= \frac{2E(SI)}{25}\left(0 + \theta_C - \frac{3(1.25\,S)}{25}\right) + 0$$

$$\Rightarrow \boxed{M_{DC} = 0.4\,EI\theta_C - 0.06\,EIS}$$

(4) Substitute step (3) into (2):

$$M_{BA} + M_{BC} = 0$$

$$\left(0.4\,EI\theta_B - 0.04\,EIS\right) + \left(0.8\,EI\theta_B + 0.4\,EI\theta_C + 0.06\,EIS\right) = 0$$

$$\Rightarrow \boxed{1.2\,EI\theta_B + 0.4\,EI\theta_C + 0.02\,EIS = 0} \quad\text{———————— (1)}$$

$$M_{CB} + M_{CD} = 0$$

$$\left(0.8\,EI\theta_C + 0.4\,EI\theta_B + 0.06\,EIS\right) + \left(0.8\,EI\theta_C - 0.06\,EIS\right) = 0$$

$$\Rightarrow \boxed{0.4\,EI\theta_B + 1.6\,EI\theta_C = 0} \quad\text{———————— (2)}$$

$$-1.333\,M_{AB} - 2.333\,M_{BA} - M_{DC} - 2\,M_{CD} = 2000$$

$$-1.333\left(0.2\,EI\theta_B - 0.04\,EIS\right) - 2.333\left(0.4\,EI\theta_B - 0.04\,EIS\right)$$

$$- \left(0.4\,EI\theta_C - 0.06\,EIS\right) - 2\left(0.8\,EI\theta_C - 0.06\,EIS\right) = 2000$$

$$\boxed{-1.2\,EI\theta_B - 2\,EI\theta_C + 0.3267\,EIS = 2000} \quad\text{———————— (3)}$$

Write the equations in matrix form:

$$EI \begin{bmatrix} 1.2 & 0.4 & 0.02 \\ 0.4 & 1.6 & 0.0 \\ -1.2 & -2.0 & 0.3267 \end{bmatrix} \begin{Bmatrix} \theta_B \\ \theta_C \\ S \end{Bmatrix} = \begin{Bmatrix} 0 \\ 0 \\ 2000 \end{Bmatrix}$$

Use program SOLVE to get the results:

$$EI\theta_B = -107.13$$
$$EI\theta_C = +26.78$$
$$EI\delta = 5892.28$$

(5) Substitute the results in step (3)

$$M_{AB} = 0.2(-107.13) - 0.04(5892.28) = -257.1 \text{ k-ft}$$

$$M_{BA} = 0.4(-107.13) - 0.04(5892.28) = -272.5 \text{ k-ft}$$

$$M_{BC} = 0.8(-107.13) + 0.4(26.78) + 0.06(5892.28)$$
$$= +272.5 \text{ k-ft}$$

$$M_{CB} = 0.8(26.78) + 0.4(-107.13) + 0.06(5892.28)$$
$$= +332.1 \text{ k-ft}$$

$$M_{CD} = 0.8(26.78) - 0.06(5892.28) = -332.1 \text{ k-ft}$$

$$M_{DC} = 0.4(26.78) - 0.06(5892.28) = -342.2 \text{ k-ft}$$

(6) Use __Statics__ to get the shear and reactions:

__NOTE__: In this problem, it is __not__ required to find the reactions, so we can draw __B.M.D.__ directly

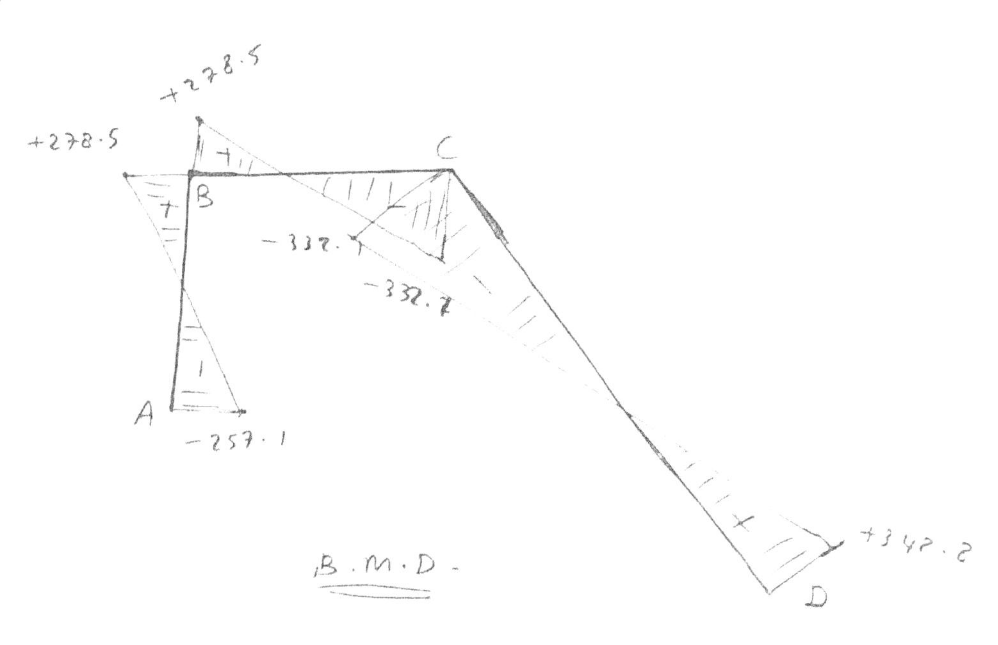

Moment—Distribution Method

14.1 Introduction:

* This is a displacement method
  (the unknowns are displacements).

* No solution of simultaneous equations is required.

* This is an <u>iteration</u> technique (with successive approximations).

* Professor Hardy Cross (University of Illinois), 1924.
  $\Rightarrow$ rapid acceptance in 1930 and 1932.

* Dominant method of analysis from the early 1930s until the emergence of matrix computer methods in the late 1950s.

$$M_{AB} = \frac{2EI}{L}\left(2\theta_A + \theta_B - \frac{3\delta}{L}\right) \pm M^F_{AB}$$

$$= \left(\frac{4EI}{L}\theta_A + \frac{2EI}{L}\theta_B\right) - \frac{6EI}{L^2}\delta \pm M^F_{AB}$$

$$M_{AB} = \underbrace{\frac{2EI}{L}(2\theta_A+\theta_B)} - \underbrace{\frac{6EI\,\delta}{L^2}} \pm \underbrace{M^F_{AB}}$$

moments due to end rotations ; moment caused by relative displacement (translation) ; fixed-end moments

---

Moment-Distribution Method:

(1) all joints are assumed to be <u>fixed</u> and the external loads are applied.

(2) Joints that can do so are permitted to rotate one at a time. (this adds moments to the fixed end moments due to joint rotation).

(3) Joints that are able to translate are allowed to do so (a second correction is added to the existing member moments).

Example :

(1) assume the beam to be initially fixed at _all_ joints.

(2) Apply a 20-k-ft (external) moment at joint B (Locking Moment)

[In order to prevent joint B from rotating].

(3) Distribute the unlocking moment at joint B.

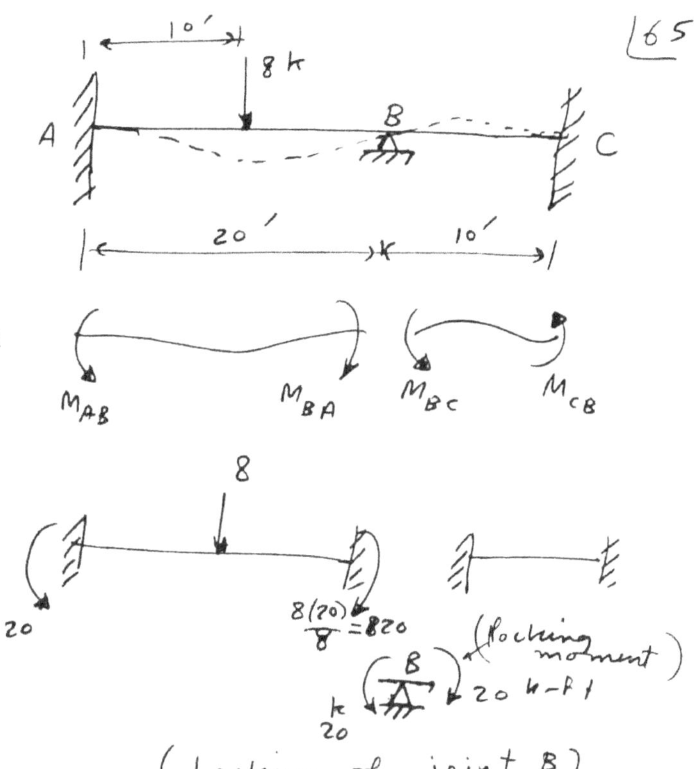

$M_{AB} = 20 + 3.33 = 23.33$ k-ft, $M_{BA} = 20 - 6.67 = 13.33$, $M_{BC} = 13.33$, $M_{CB} = 6.67$

14.2 Basic Concepts :

1. Sign Convention.
2. Fixed-end Moments.
3. Distribution Factors.
4. Carry-over Moments

Sign Convention:

* the same sign convention used in the slope-deflection method is used in the moment-distribution method.

⟹ the moment acting on the end of a member is positive when it acts clockwise, and negative when it acts counterclockwise.

Fixed-End Moments:

* these are the moments that develop at the ends of a member as a result of the loads acting along the member if both ends of the member are fixed.
* Members having no loads acting along their span have zero fixed-end moments.

Distribution Factors:

* Joints are alternately <u>locked</u> and <u>unlocked</u>.
* When the joints are unlocked and permitted to rotate, the rotation is resisted by the members that frame into the joint.
* the term <u>distribution factor</u> signifies that the resistance to the rotation of a joint is distributed among the members framing into the joint.
* Specifically, the distribution factor for any one member is a measure of the proportion of the total resistance to rotation supplied by that member.

Example:

A: rigid joint.

(All joints are fixed except the one that is being permitted to rotate).

$\Rightarrow$ three resisting moments will develop.

$$M_{AB} = \frac{2EI}{L}\left(2\theta_A + \theta_B - 3\frac{\delta}{L}\right) \pm M_{AB}^F$$

$$= \frac{2EI}{L}(2\theta + 0 - 0) \pm 0$$

$$\Rightarrow M_{AB} = \frac{4EI\theta}{L} \Rightarrow \begin{cases} M_{AB} = \dfrac{4E\,I_{AB}\,\theta}{L_{AB}} \\[2mm] M_{AC} = \dfrac{4E\,I_{AC}\,\theta}{L_{AC}} \\[2mm] M_{AD} = \dfrac{4E\,I_{AD}\,\theta}{L_{AD}} \end{cases}$$

Similarly,

Define the _stiffness_ k of a member as follows:

Member _Stiffness_: $\quad k = \dfrac{I}{L} \qquad \left(\dfrac{m^4}{m} = m^3 \right)$.

$\Longrightarrow \qquad M_{AB} = 4E\theta\, k_{AB}$

$\qquad\qquad M_{AC} = 4E\theta\, k_{AC}$

$\qquad\qquad M_{AD} = 4E\theta\, k_{AD}$

At joint A : $+\circlearrowleft \Sigma M_A = 0$: $\quad M_{AB} + M_{AC} + M_{AD} - M = 0$

$\Longrightarrow \qquad M = M_{AB} + M_{AC} + M_{AD}$

$\Longrightarrow \qquad M = 4E\theta\, k_{AB} + 4E\theta\, k_{AC} + 4E\theta\, k_{AD}$

$\qquad\qquad\quad = 4E\theta \left(k_{AB} + k_{AC} + k_{AD} \right)$

$\Longrightarrow \qquad M = 4E\theta\, \Sigma k_A$

where Σk_A : sum of the stiffnesses of the members framing into joint A.

$\qquad\qquad 4E\theta = \dfrac{M}{\Sigma k_A}$

$\Longrightarrow \qquad M_{AB} = 4E\theta\, k_{AB} = \dfrac{M}{\Sigma k_A} \cdot k_{AB} = \left(\dfrac{k_{AB}}{\Sigma k_A} \right) M$

$\qquad\qquad M_{AC} = \left(\dfrac{k_{AC}}{\Sigma k_A} \right) M$

$\qquad\qquad M_{AD} = \left(\dfrac{k_{AD}}{\Sigma k_A} \right) M$

$\dfrac{k}{\Sigma k} =$ distribution factor of the member.

$\qquad\qquad \hookrightarrow$ this gives the fraction of the total moment applied to a joint that is resisted by that member.

Carry-Over Moments:

* the resulting moment M_{DA} at the other end of the member M_{DA} is called the "carry-over moment"

* the ratio $\dfrac{M_{DA}}{M_{AD}}$ = carry-over factor.

$$M_{DA} = \frac{2EI}{L}\left(2\theta_D + \theta_A - \frac{3\delta}{L}\right) \pm M_{DA}^F$$

$$= \frac{2EI}{L}\left(0 + \theta_A - 0\right) \pm 0$$

$$= \frac{2EI\theta_A}{L}$$

$$= \frac{1}{2} M_{AD} \implies \boxed{M_{DA} = \frac{1}{2} M_{AD}}$$

* Carry-Over Factor $\equiv \left(\dfrac{1}{2}\right)$: is the ratio of the moment induced at the fixed end to the moment applied at that end that is able to rotate,

---

14.3 Application of Moment Distribution to Structures Without Joint Translation:

Example 14.1:

Determine the member moments and plot the moment diagram for the beams shown in the figure.

Solution:

(1) Calculate the distribution factors at all joints that are free to rotate.

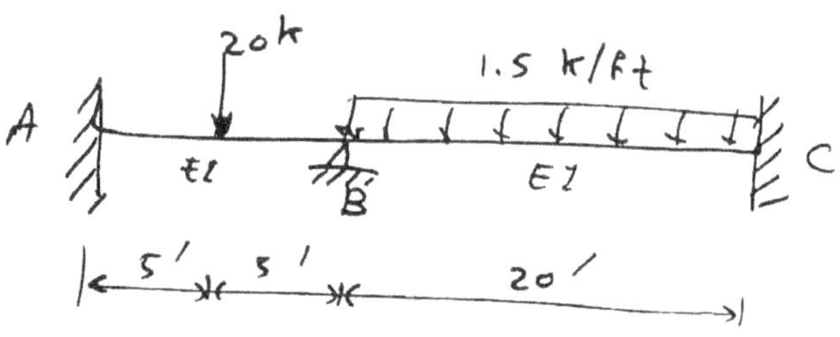

At joint B:

$$\Sigma k_B = \frac{I}{10} + \frac{I}{20} = \frac{3I}{20}$$

$$DF_{AB} = \frac{k_{AB}}{\Sigma k_B} = \frac{\frac{I}{10}}{\frac{3I}{20}} = \frac{2}{3}$$

$$DF_{BC} = \frac{k_{BC}}{\Sigma k_B} = \frac{\frac{1}{20}}{\frac{3}{20}} = \frac{1}{3}.$$

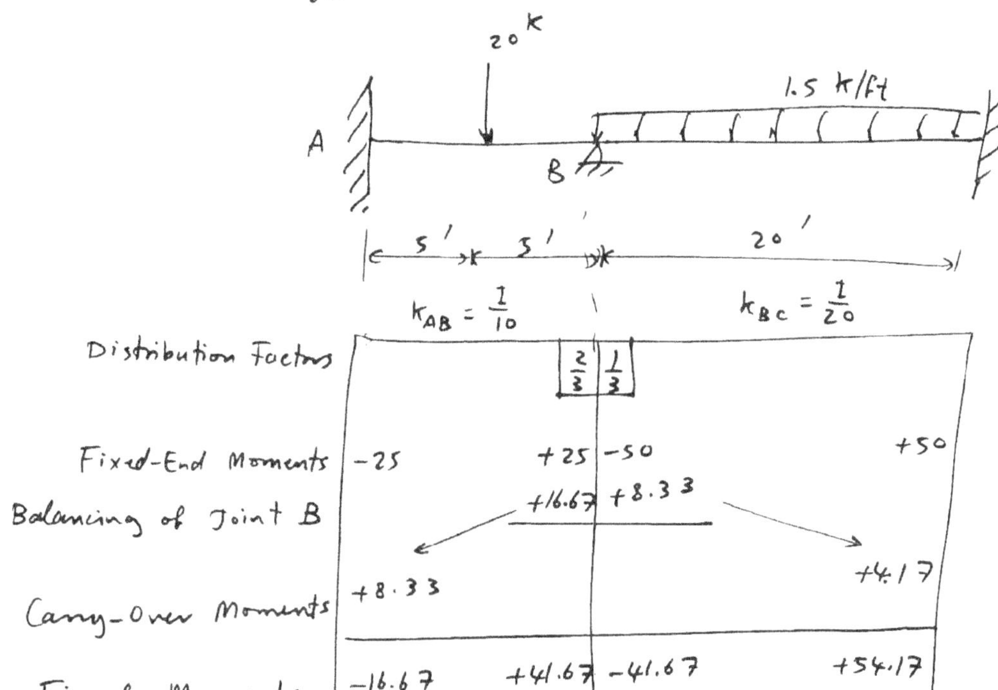

(2) Assume all joints to be locked and calculate the fixed-end moments.

$$M^F_{AB} = -\frac{PL}{8} = -\frac{20(10)}{8} = -25 \; k\text{-}ft.$$

$$M^F_{BA} = +\frac{PL}{8} = +25 \; k\text{-}ft.$$

$$M^F_{BC} = -\frac{wL^2}{12} = -\frac{(1.5)(20)^2}{12} = -50 \; k\text{-}ft.$$

$$M^F_{CB} = +\frac{wL^2}{12} = +50 \; k\text{-}ft.$$

(3) Begin the actual moment-distribution process. (an external moment of 25 k-ft was needed to lock joint B). The unlocking moment will be distributed among the two members in accordance with the distribution factors.

(4) Carry-over moments are calculated at the other (fixed) ends of the members.

(5) Joints A and C are fixed, and joint B is externally unrestrained ⟹ the moment-distribution process is complete.

(6) Final moments are calculated:

$$M_{AB} = -16.67 \text{ k-ft}.$$
$$M_{BA} = +41.67 \text{ k-ft}.$$
$$M_{BC} = -41.67 \text{ k-ft}.$$
$$M_{CB} = +54.17 \text{ k-ft}.$$

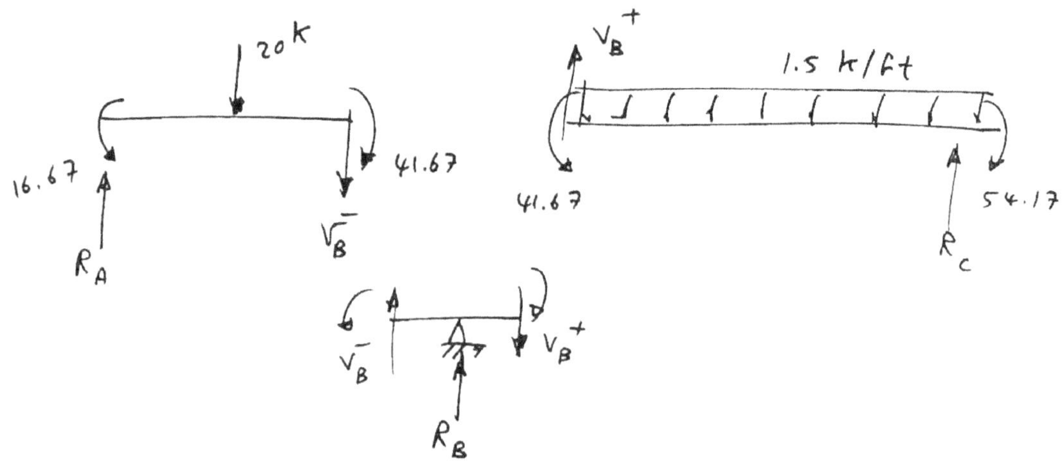

Member AB : $+\circlearrowleft \Sigma M_A = 0$:

$$16.67 - 41.67 - V_B^- (10) - 20(5) = 0$$
$$V_B^- = -12.5 \text{ k}.$$

$+\uparrow \Sigma F_y = 0$: $R_A - 20 - (-12.5) = 0 \implies R_A = +7.5 \text{ k} \uparrow$

Member BC: $+\circlearrowleft \Sigma M_B = 0$:

$$R_c(20) - 1.5(20)(10) - 54.17 + 41.67 = 0$$
$$\implies R_c = 15.625 \text{ k} \uparrow.$$

$+\uparrow \Sigma F_y = 0$:

$$V_B^+ + 15.625 - 1.5(20) = 0$$
$$\implies V_B^+ = 14.375 \text{ k}.$$

Joint B : $R_B = 14.375 - (-12.5) = 26.88 \text{ k} \uparrow.$

Locate point 0 of zero shear:

$$\frac{x}{14.38} = \frac{20-x}{15.62}$$
$$15.62x = 20(14.38) - 14.38x$$
$$x = 9.59'.$$

$$M_0 = -41.67 + \frac{1}{2}(14.38)(9.59)$$
$$= 27.28 \text{ k-ft}$$

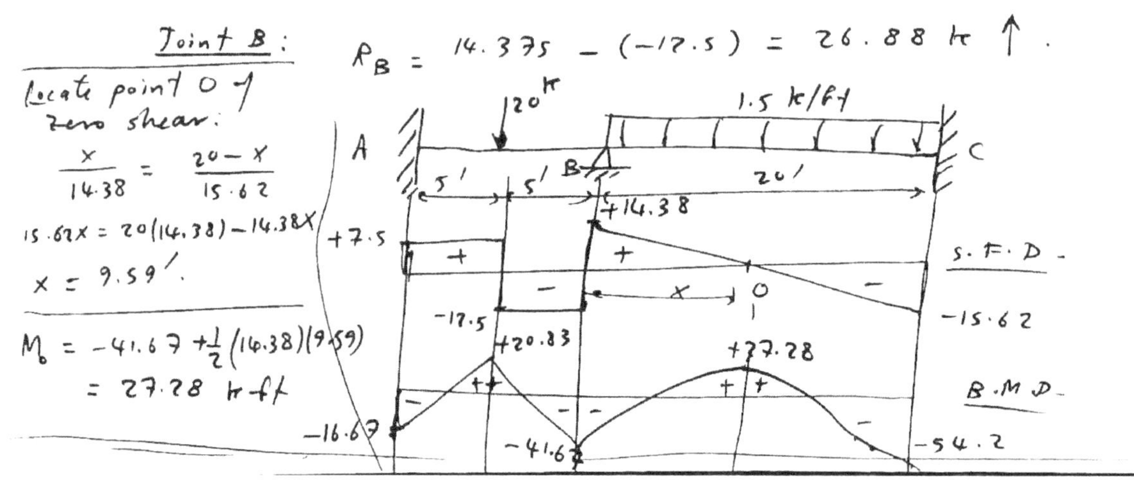

Example 14.2 :

Determine the member moments and draw the moment diagram for the structure shown in the figure.

Solution :

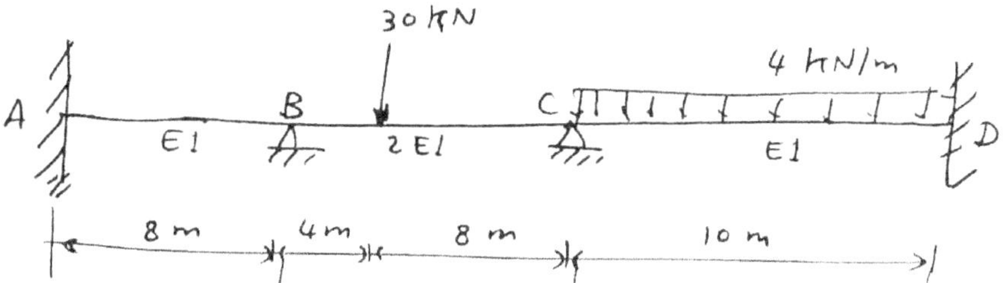

Joint B :

$\Sigma k_B = \dfrac{I}{8} + \dfrac{I}{6}$

$\quad = \dfrac{6I + 8I}{48}$

$\quad = \dfrac{14I}{48} = \dfrac{7I}{24}$

$\Sigma k_C = \dfrac{I}{6} + \dfrac{I}{10}$

$\quad = \dfrac{10I + 6I}{60}$

$\quad = \dfrac{16I}{60} = \dfrac{4I}{15}$

$DF_{AB} = \dfrac{\frac{I}{8}}{\frac{7I}{24}} = \dfrac{24^{3}}{8 \times 7}$

$\quad = \dfrac{3}{7}$

$DF_{BC} = \dfrac{\frac{I}{6}}{\frac{7I}{24}} = \dfrac{24^{4}}{6 \times 7}$

$\quad = \dfrac{4}{7}$

$DF_{CB} = \dfrac{I/6}{4I/15} = \dfrac{15^{5}}{6 \times 4}$

$\quad = \dfrac{5}{8}$

$DF_{CD} = \dfrac{I/10}{4I/15} = \dfrac{15^{3}}{4 \times 10}$

$\quad = \dfrac{3}{8}$

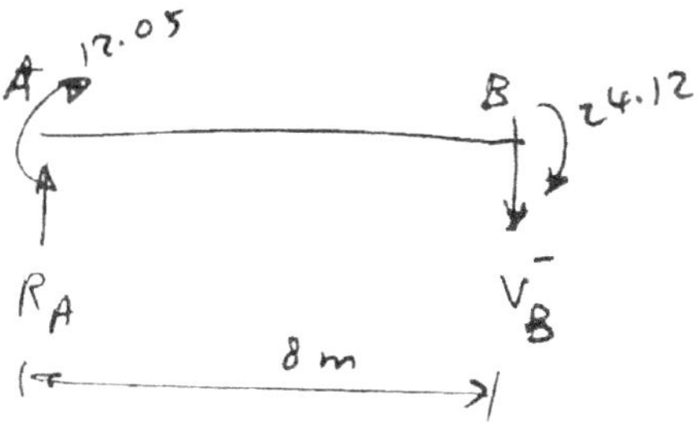

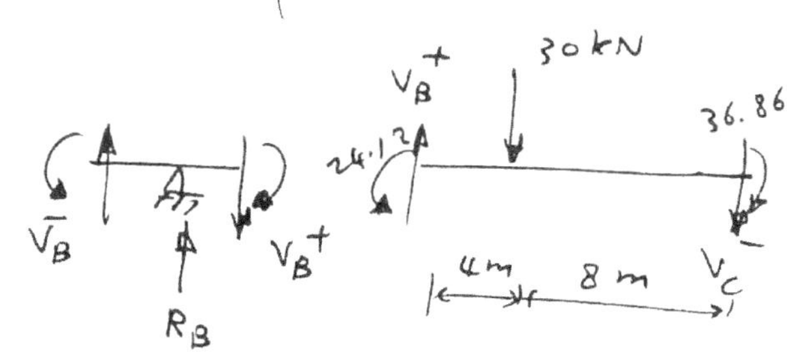

Moment distribution table (upper figure):

Member	A	B		C		D
k		$k=\frac{I}{8}$	$k=\frac{2I}{12}=\frac{I}{6}$		$k=\frac{I}{10}$	
DF		$\frac{3}{7}$	$\frac{4}{7}$	$\frac{5}{8}$	$\frac{3}{8}$	
FEM		−53.33	+26.67	−33.33	+33.33	
	+11.43	+22.86 +30.47		+15.24		
			−5.36	−3.22	−1.61	
		−2.68				
	+0.57	+1.15 +1.53		+0.76		
			−0.48	−0.29	−0.14	
		−0.24				
	+0.05	+0.10 +0.14		+0.07		
			−0.04	−0.03	−0.015	
		−0.02				
		+0.01 +0.01				
Total	+12.05	+24.12 −24.12	+36.86	−36.87	+31.56	

Dimensions: 8 m, 4 m, 8 m, 10 m

Loads: 30 kN at B–C span; 4 kN/m UDL on C–D span.

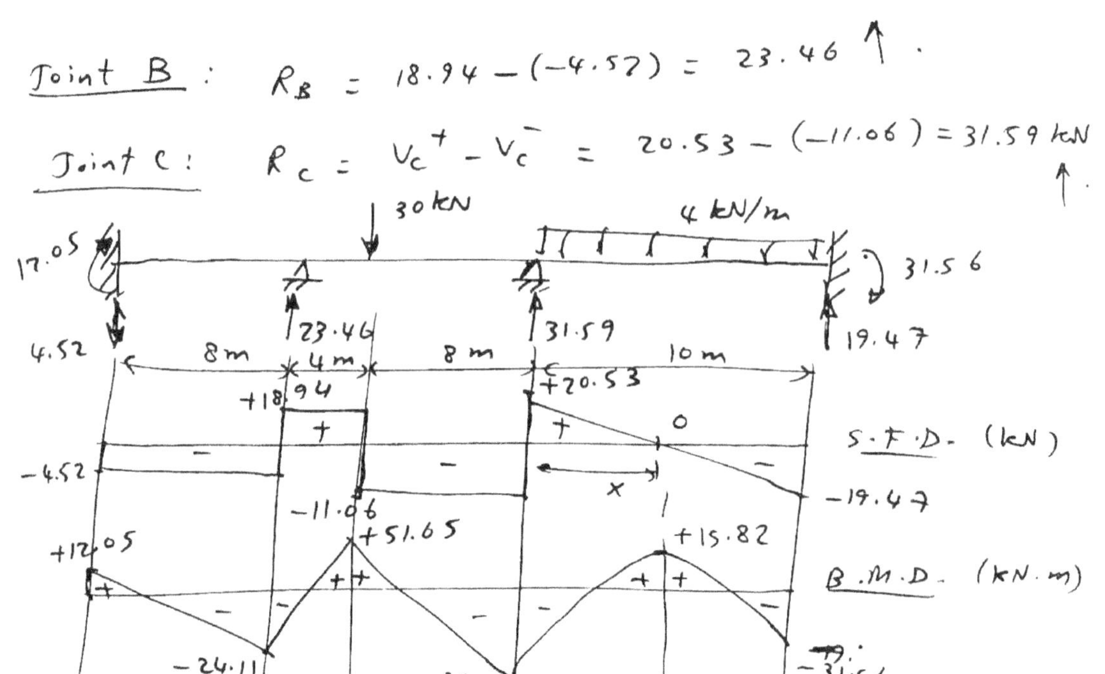

$$V_c^- \quad R_c \quad V_c^+$$

$$36.86 \quad 4\,kN/m \quad 31.56 \quad R_D \quad 10\,m$$

Member AB : $+\circlearrowleft \; \Sigma M_A = 0 :\quad -V_B^-(8) - 24.12 - 12.05 = 0$

$$V_B^- = -4.52 \; kN .$$

$\qquad +\uparrow \; \Sigma F_y = 0 : \qquad R_A - (-4.52) = 0$

$$R_A = -4.52 \; kN$$
$$R_A = 4.52 \; kN \downarrow .$$

Member BC :

$+\uparrow \; \Sigma M_B = 0 : \qquad -V_c^-(12) - 36.86 - 30(4) + 24.12 = 0$

$$V_c^- = -11.06 \; kN .$$

$+\uparrow \; \Sigma F_y = 0 : \qquad V_B^+ - (-11.06) - 30 = 0$

$$V_B^+ = 18.94 \; kN .$$

Member CD :

$+\circlearrowleft \; \Sigma M_c = 0 : \qquad R_D(10) - 31.56 - 4(10)(5) + 36.86 = 0$

$$R_D = 19.47 \; kN \uparrow .$$

$+\uparrow \; \Sigma F_y = 0 : \qquad V_c^+ + 19.47 - 4(10) = 0$

$$V_c^+ = 20.53 \; kN .$$

Joint B : $\quad R_B = 18.94 - (-4.52) = 23.46 \uparrow .$

Joint C : $\quad R_c = V_c^+ - V_c^- = 20.53 - (-11.06) = 31.59 \; kN \uparrow$

Locate point O of zero shear:

$$\frac{x}{20.53} = \frac{10-x}{19.47}$$

$$19.47x = 10(20.53) - 20.53x \implies x = 5.13 \text{ m}.$$

$$M_0 = -36.83 + \frac{1}{2}(5.13)(20.53) = 15.82 \text{ kN}\cdot\text{m}.$$

$\underline{Example\ 14.3}$:

Determine the member moments for the two-span beam shown in the figure. The member is fixed at A, continuous over the support at B, and simply-supported at C.

$\underline{Solution}$:

$D.F._{BC}$ at joint $C = 1$

At joint B :

$$\Sigma k = \frac{I}{20} + \frac{2I}{20} = \frac{3I}{20}$$

$$D.F._{BA} = \frac{I/20}{3I/20} = \frac{1}{3}$$

$$D.F._{BC} = \frac{2}{3}.$$

$$M_A = +3.0 \ k-ft.$$
$$\therefore \ M_B = -6.0 \ k-ft.$$

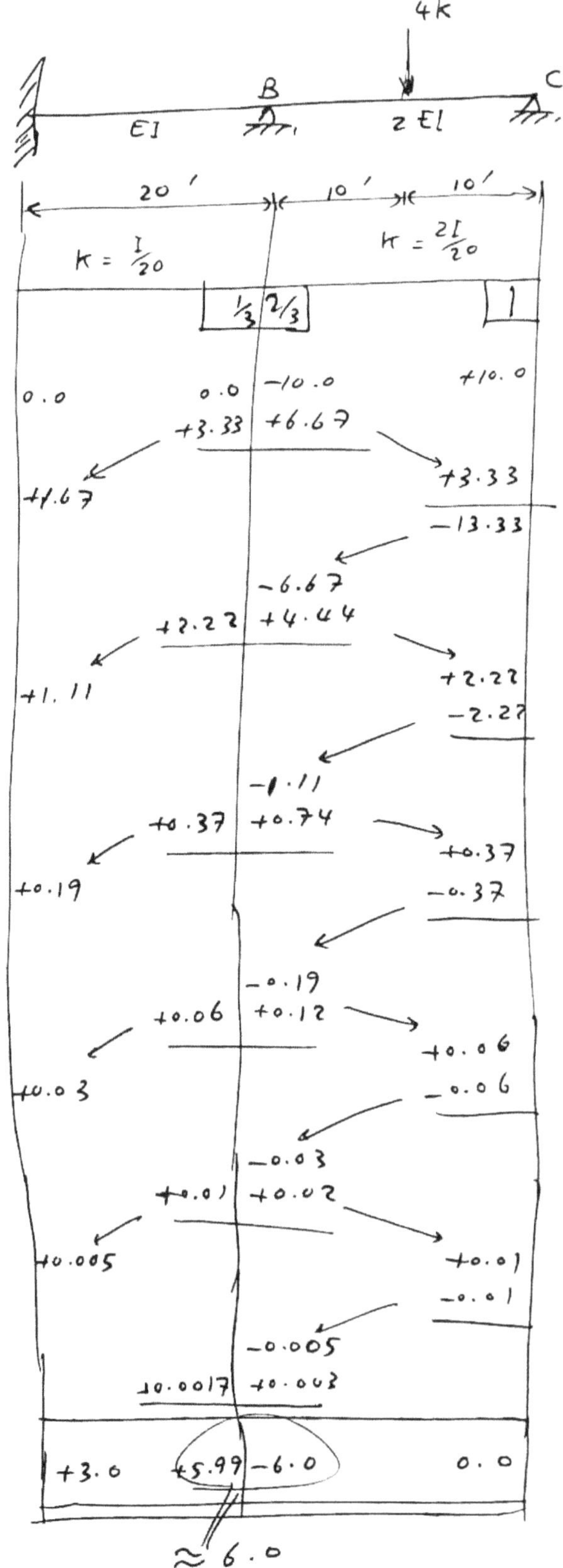

Simply-Supported Ends :

* Introduce certain modifications in the moment-distribution procedure to reduce the numerical work.

$$M_{AB} = 4Ek_{AB}\,\theta_A$$

$$M_{AC} = 4Ek_{AC}\,\theta_A$$

$$M_{AD} = 4Ek_{AD}\left(\theta_A + \tfrac{1}{2}\theta_D\right)$$

But

$$M_{DA} = 4Ek_{AD}\left(\theta_D + \tfrac{1}{2}\theta_A\right) = 0$$

$$\implies \quad \theta_D = -\tfrac{1}{2}\theta_A$$

$$\implies \quad M_{AD} = 4E\,k_{AD}\left(\theta_A - \tfrac{1}{2}\left(\tfrac{1}{2}\theta_A\right)\right)$$

$$= 4Ek_{AD}\left(\tfrac{3}{4}\theta_A\right)$$

$$\implies \quad \boxed{M_{AD} = 4E\theta_A\left(\tfrac{3}{4}k_{AD}\right)}$$

$$+\!\!\circlearrowleft \;\; \Sigma M_A = 0 \implies \quad M = M_{AB} + M_{AC} + M_{AD}$$

$$M = 4Ek_{AB}\,\theta_A + 4Ek_{AC}\,\theta_A + 4E\theta_A\left(\tfrac{3}{4}k_{AD}\right)$$

$$= 4E\theta_A\left(k_{AB} + k_{AC} + \tfrac{3}{4}k_{AD}\right)$$

$$M = 4E\theta_A\,\Sigma k'_A$$

where $k' \equiv$ effective stiffness of a member.

$$k' = \begin{cases} I/L & , \text{ if the member is } \underline{\text{fixed}} \text{ at its far end.} \\ 0.75\,I/L & , \text{ if the member is } \underline{\text{hinged}} \text{ at its far end.} \end{cases}$$

$$4E\theta_A = \frac{M}{\Sigma k'_A}$$

$$\Rightarrow \quad M_{AB} = \left(\frac{k'_{AB}}{\Sigma k'_A}\right) M$$

$$M_{AC} = \left(\frac{k'_{AC}}{\Sigma k'_A}\right) M$$

$$M_{AD} = \left(\frac{k'_{AD}}{\Sigma k'_A}\right) M$$

<u>Note</u> : ① Modified D.F. $= \dfrac{k'}{\Sigma k'}$.

② <u>No</u> moment is induced at D.

$\Rightarrow$ Carry-Over Moment to a <u>Hinged End</u> is <u>zero</u>.

<u>Example 14.4</u> :

Determine the member moments for the beam shown in the figure using effective stiffnesses.

<u>Solution</u> :

$$\Sigma k'_B = \frac{1}{20} + \frac{3}{4}\left(\frac{2I}{20}\right)$$

$$= \left(\frac{4+6}{4(20)}\right) I$$

$$= \frac{1}{8} I \ .$$

$$\frac{k'_{AB}}{\Sigma k'_B} = \frac{1/20}{1/8} = \frac{8}{20} = \frac{2}{5} \ .$$

$$\frac{k'_{BC}}{\Sigma k'_B} = \frac{\frac{3}{4}\left(\frac{2I}{20}\right)}{1/8} = \frac{3}{4}\cdot\frac{8}{10} = \frac{3}{5} \ .$$

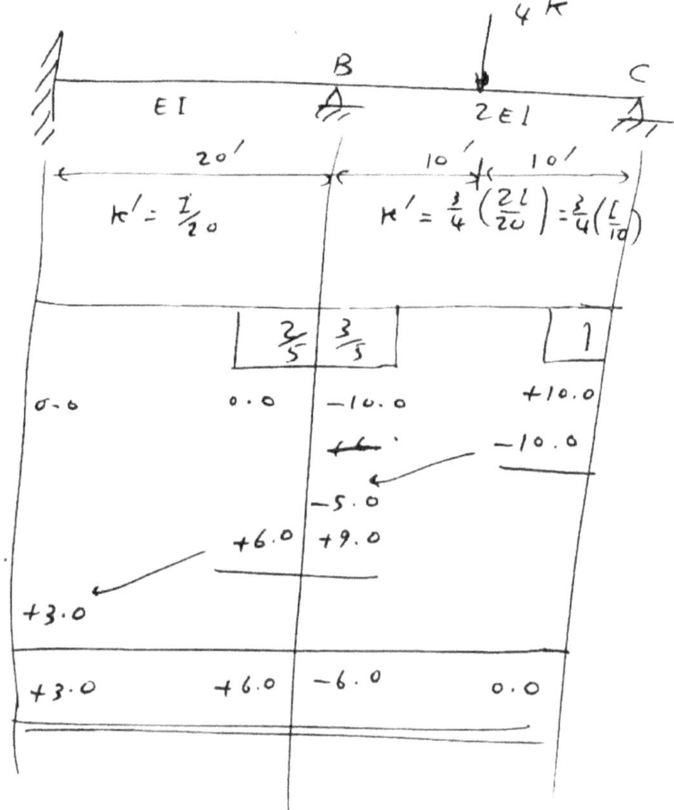

<u>Note</u> :
Unlock all joints at simple-supports first.

Example 14.5 :

Determine the member moments for the structure shown in the figure.

Solution :

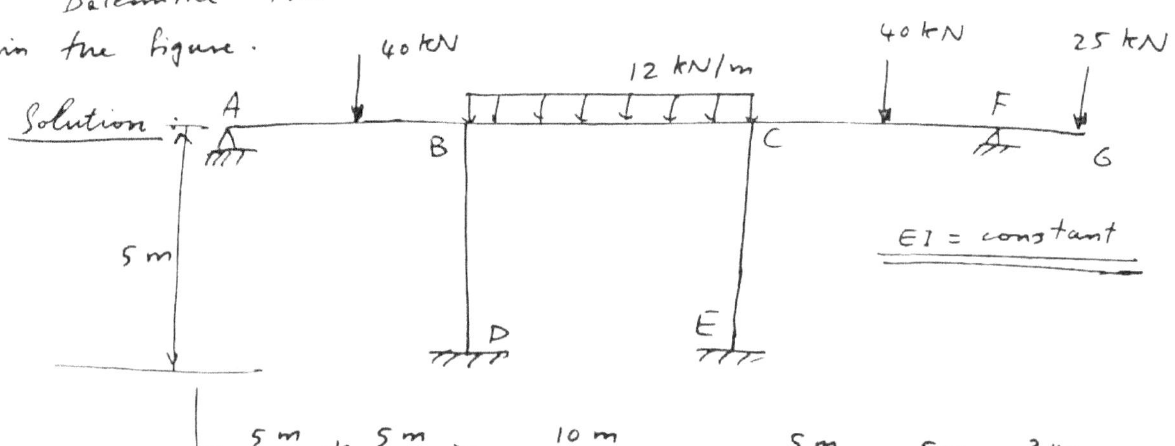

Joint B:

$$\Sigma k'_B = \frac{3}{4}\left(\frac{I}{10}\right) + \frac{I}{10} + \frac{I}{5}$$

$$= \frac{3I + 4I + 8I}{4(10)}$$

$$= \frac{15I}{40} = \frac{3}{8}I$$

Joint C:

$$\Sigma k'_C = \frac{I}{10} + \frac{I}{5} + \frac{3}{4}\left(\frac{I}{10}\right) = \frac{3I}{8}$$

$$\text{D.F.}_{AB} = \frac{\frac{1}{4}\left(\frac{I}{10}\right)}{\frac{3}{8}I} = \frac{3}{4} \cdot \frac{1}{10} \cdot \frac{8}{3} = \frac{1}{5} = \frac{3}{15}$$

$$\text{D.F.}_{BC} = \frac{I/10}{\frac{3}{8}I} = \frac{1}{10} \cdot \frac{8}{3} = \frac{4}{15}$$

$$\text{D.F.}_{BD} = \frac{I/5}{3\frac{I}{8}} = \frac{1}{5} \cdot \frac{8}{3} = \frac{8}{15}$$

14.4 Structures with Joint Translations :
(Frames with Sidesway)

* For structures with joint translations, we need to carry out the moment distribution method in two stages.

Example :

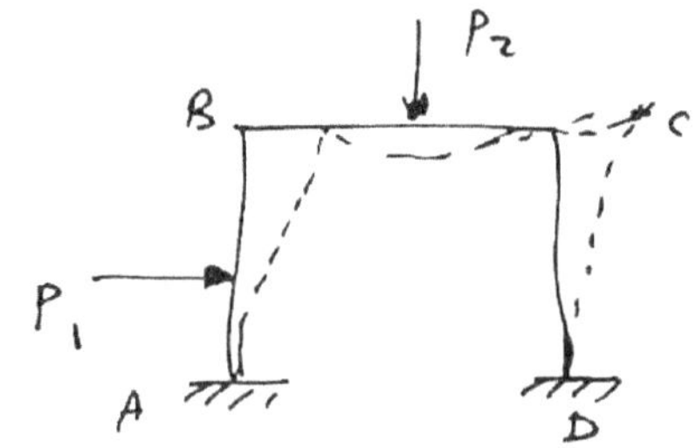

∴ Joints B and C translate in the above frame.

(1) Add a sufficient number of artificial restraints to the structure so that no joint translations occur.

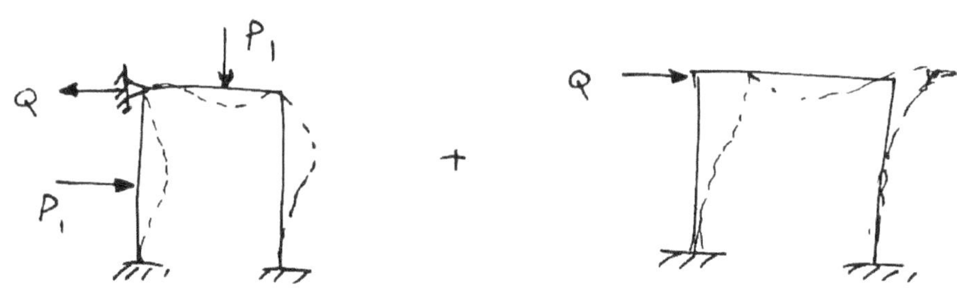

(2) A routine moment distribution is then carried out for the artificially restrained structure.

(3) A second set of calculations, whose purpose is to cancel out the effects of the artificial restraint is then performed.

(4) The member moments thus produced are added to those obtained in the previous stage of the analysis.

Example 14.6 :

Determine the member moments for the structure shown in the figure.

Solution :

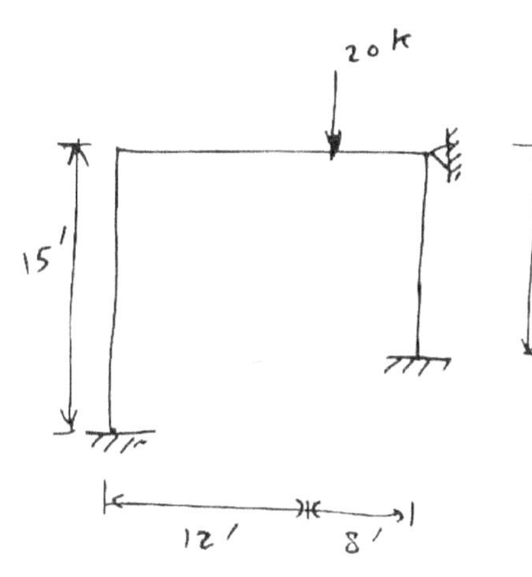

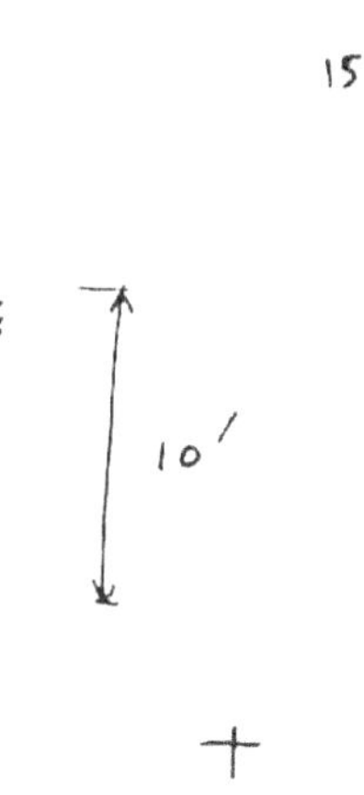

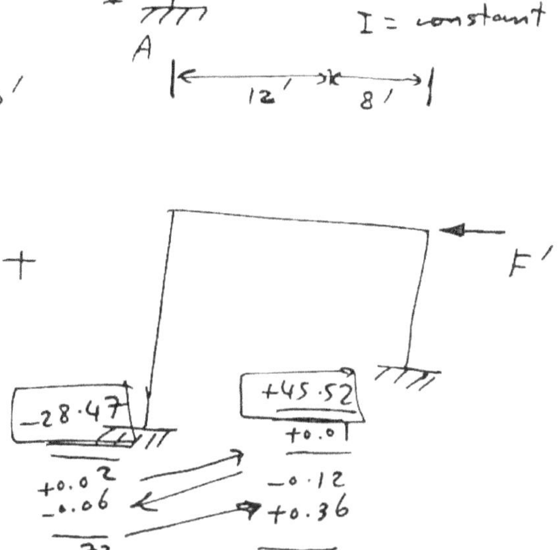

Joint B :

$$\Sigma K_B = \frac{I}{15} + \frac{I}{20} = 0.1167\,I$$

$$DF_{AB} = \frac{I/15}{0.1167\,I} = 0.571$$

$$DF_{BC} = \frac{I/20}{0.1167\,I} = 0.428$$

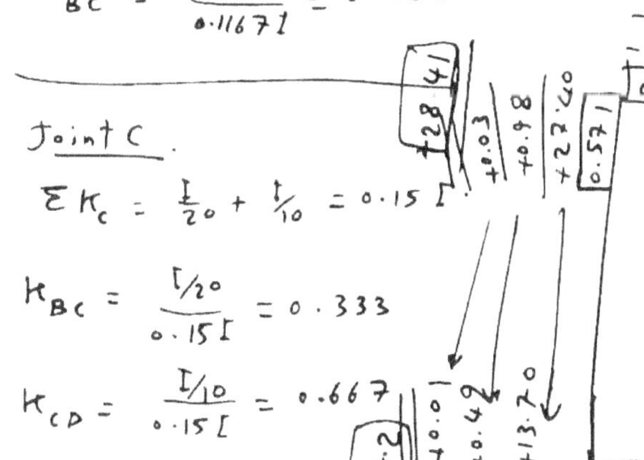

Joint C .

$$\Sigma K_c = \frac{I}{20} + \frac{I}{10} = 0.15\,I$$

$$K_{BC} = \frac{I/20}{0.15\,I} = 0.333$$

$$K_{CD} = \frac{I/10}{0.15\,I} = 0.667$$

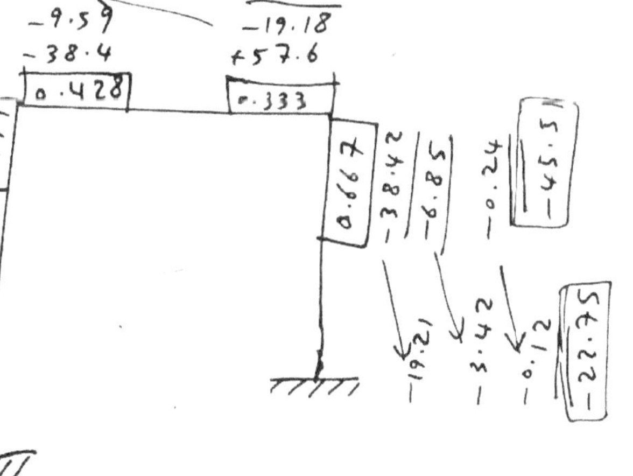

* Determine the magnitude of the restraining force :

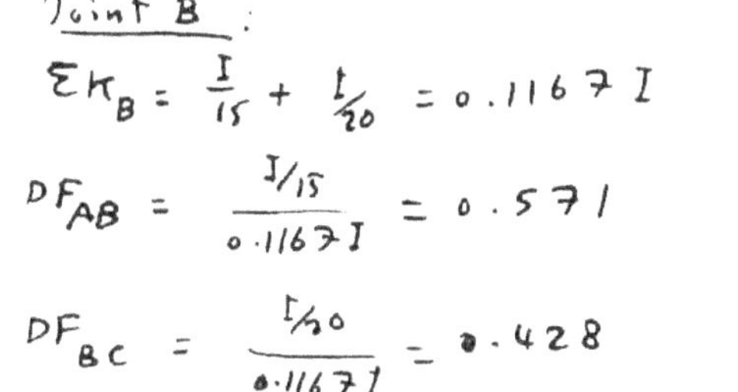

$$H_A = \frac{28.41 + 14.2}{15} = 2.84\ k .$$

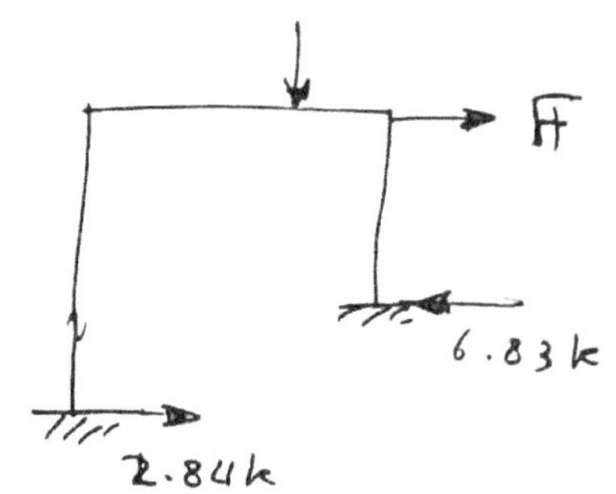

$$H_D = \frac{45.31 + 22.75}{10} = 6.83 \text{ k}.$$

$$\underline{\Sigma F_x = 0}:$$

$$F + 2.84 - 6.83 = 0$$

$$F = 3.99 \text{ k} \longrightarrow .$$

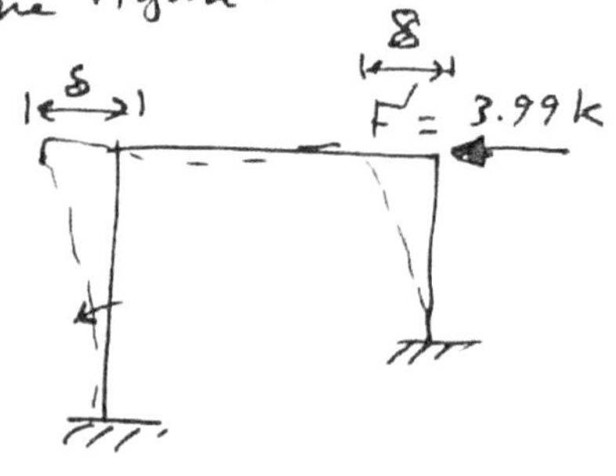

Apply a force F', which is equal and opposite to F, to the structure as shown in the figure.

End moments in members AB and CD are given by

$$M_{AB} = \frac{6EI\delta}{L^2}.$$

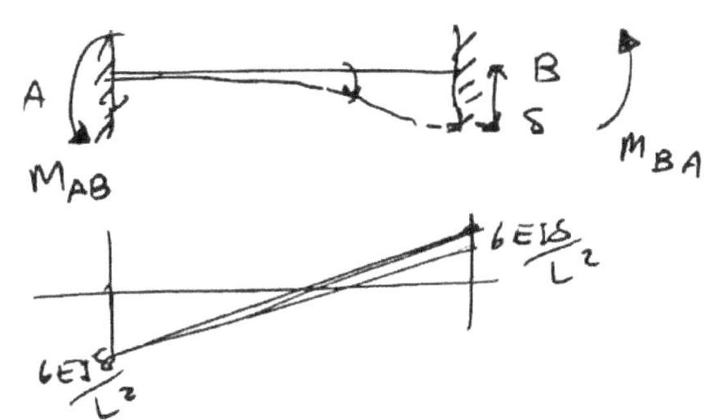

$$M_{AB} = + \frac{6EI\delta}{L^2} \qquad\qquad M_{CD} = \frac{6EI\delta}{L^2}$$

$$M_{BA} = + \frac{6EI\delta}{L^2} \qquad\qquad M_{DC} = \frac{6EI\delta}{L^2}.$$

We do not know δ, so we take the member moments in proportions to their $1/L^2$ values.

$$\underline{AB}: \quad \frac{1}{L^2} = \frac{1}{(15)^2} = \frac{1}{225}.$$

$$\underline{CD}: \quad \frac{1}{L^2} = \frac{1}{(10)^2} = \frac{1}{100}.$$

Take $\quad M_{AB} = M_{BA} = \dfrac{1000}{225} = 4.44$

$\quad M_{CD} = M_{DC} = \dfrac{1000}{100} = 10.0$

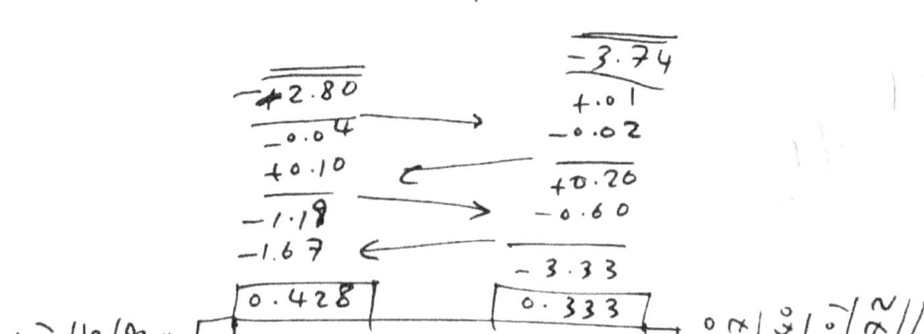

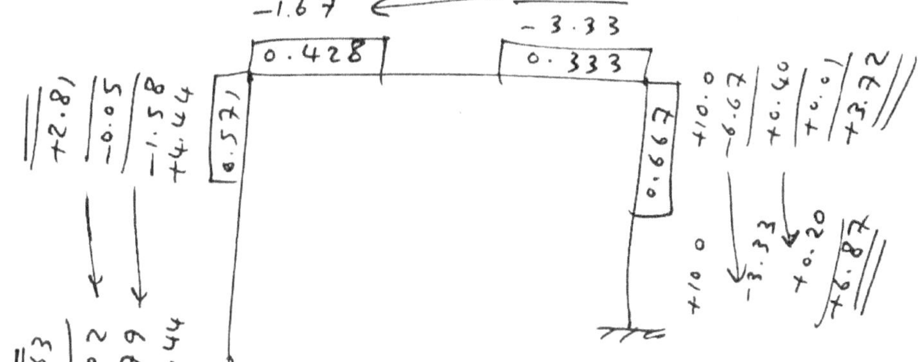

Find the force F' corresponding to these moments:

$H_A = \dfrac{3.63 + 2.81}{15} = +0.43\ k.$

$H_D = \dfrac{3.72 + 6.87}{10} = +1.06.$

$\Longrightarrow \ \Sigma F_x = 0: \quad F' = H_A + H_D$

$\qquad = 0.43 + 1.06 = +1.49\ k.$

the ratio of the two F's is $\dfrac{3.99}{1.49} = +\underline{\underline{2.68}}$.

82

By superposition:

$\therefore\ M_{AB} = +14.2 + 2.68(3.63) = +23.93$ k-ft.

$M_{BA} = +28.41 + 2.68(+2.81) = +35.94$ k-ft.

$M_{BC} = -28.47 + 2.68(-2.80) = -35.97$ k-ft.

$M_{CB} = +45.52 + 2.68(-3.74) = +35.5$ k-ft.

$M_{CD} = -45.51 + 2.68(+3.72) = -35.54$ k-ft.

$M_{DC} = -22.75 + 2.68(+6.87) = -4.34$ k-ft.

Homework 5

14.2 , $\underline{14.8}$, $\underline{14.14}$, $\underline{14.16}$.

Frames with Sloping Members:

Example:

Compute the joint moments in the frame shown in the figure using the moment distribution method.

$I = 900$ for all members.

Solution:

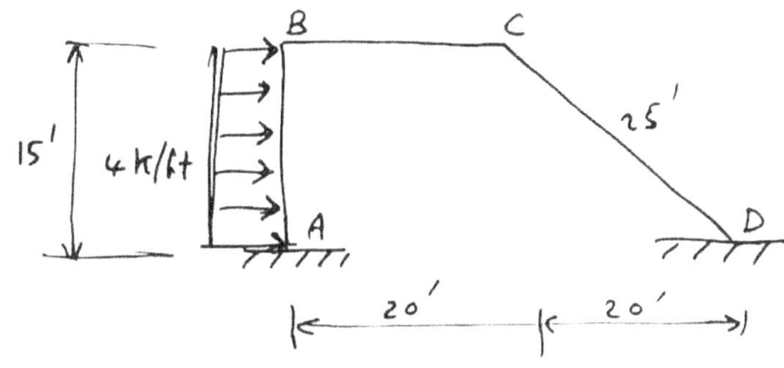

Step 1: Analyze the restrained frame (without sidesway):

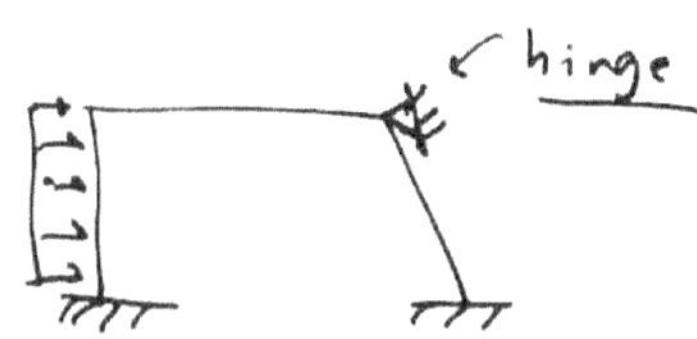

Joint B: $\Sigma k_B = \dfrac{I}{15} + \dfrac{I}{20} = 0.1167\,I$

$DF_{AB} = \dfrac{I/15}{0.1167\,I} = 0.571$

$DF_{BC} = \dfrac{I/20}{0.1167\,I} = 0.428$

Joint C: $\Sigma k_c = \dfrac{I}{20} + \dfrac{I}{25} = 0.09\,I$

$DF_{BC} = \dfrac{I/20}{0.09\,I} = 0.556$

$DF_{CD} = \dfrac{I/25}{0.09\,I} = 0.444$

Fixed-End Moments:

$M_{F_{AB}} = -\dfrac{wL^2}{12} = -\dfrac{4(15)^2}{12} = -75 \ \text{k-ft}$

$M_{F_{BA}} = +\dfrac{wL^2}{12} = +75 \ \text{k-ft}$

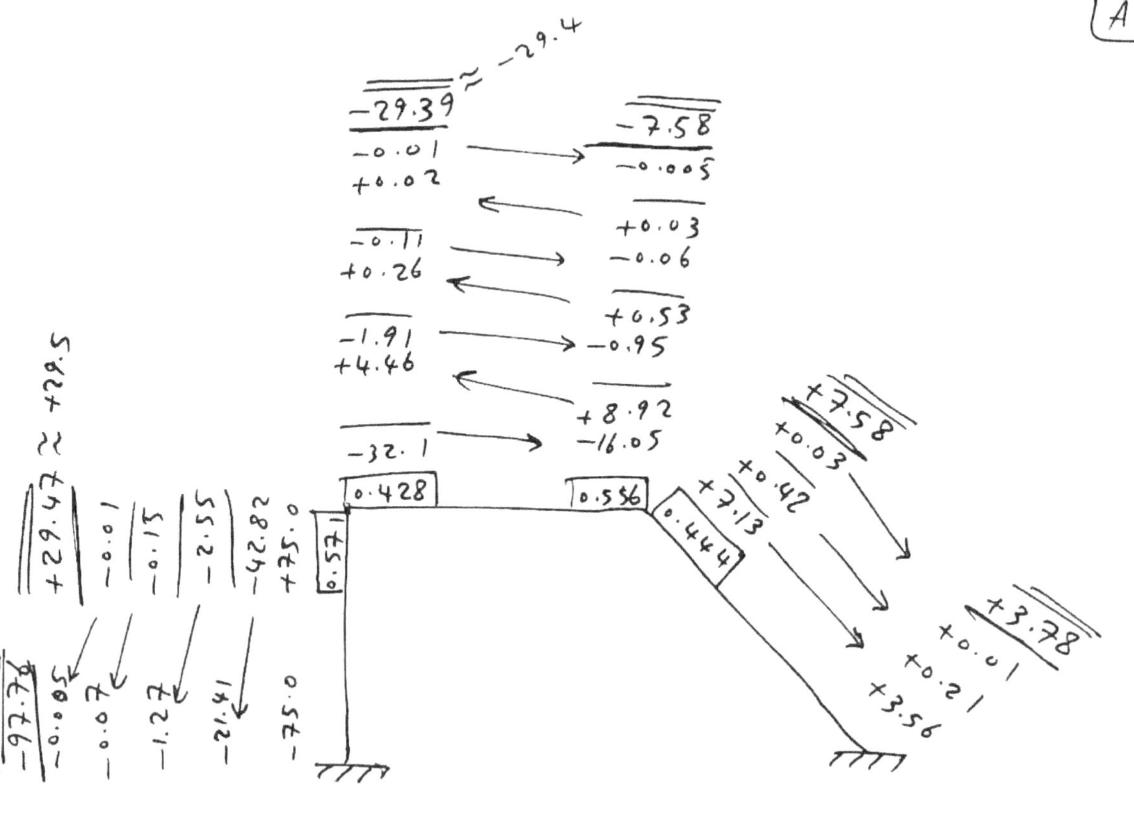

$$+\circlearrowleft \ \Sigma M_B = 0:$$

$$-29.47 + 97.76 + 4(15)\left(\frac{15}{2}\right) - H_A(15) = 0$$

$$H_A = +34.55^{\ k}$$

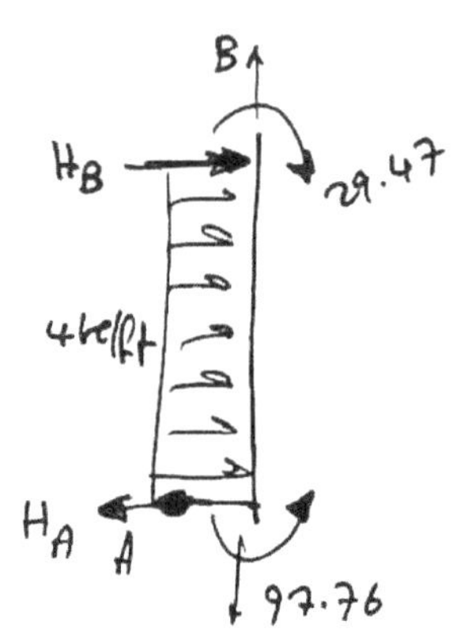

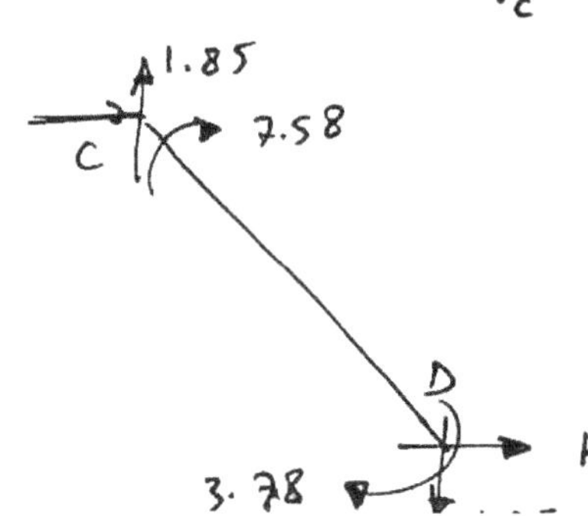

$$V_C = \frac{7.58 + 29.4}{20} = 1.85 \ k.$$

$$+\circlearrowleft \ \Sigma M_C = 0:$$

$$-7.58 - 3.78 - 1.85(20) + H_D(15) = 0$$

$$H_D = 3.22 \ k$$

For the entire frame.

$\xrightarrow{+} \Sigma F_x = 0:$

$F = -34.55 + 3.22 + 4(15)$

$\therefore \quad F = +28.67 \leftarrow$

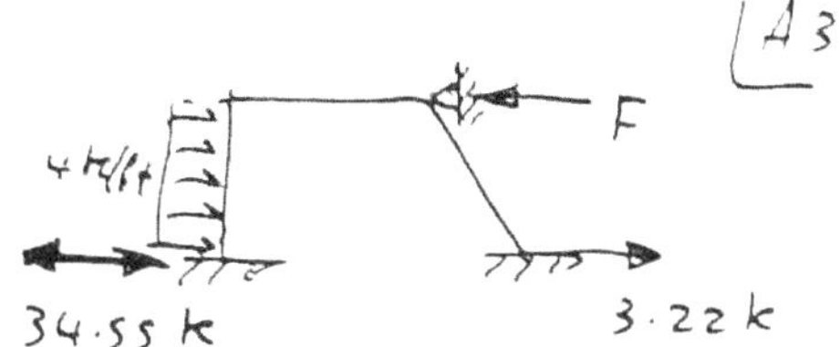

Step 2 : Analyze the frame with sidesway ONLY.

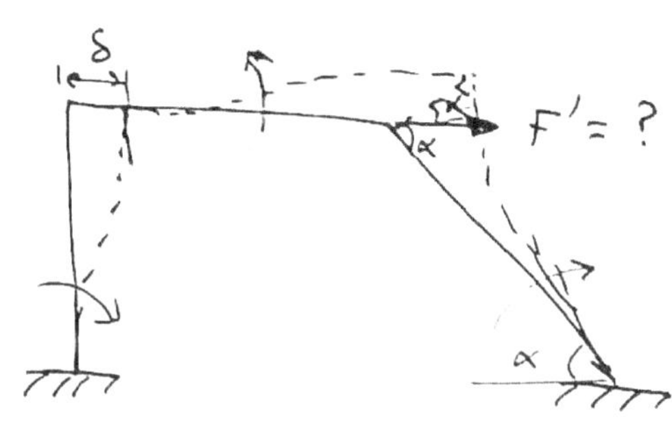

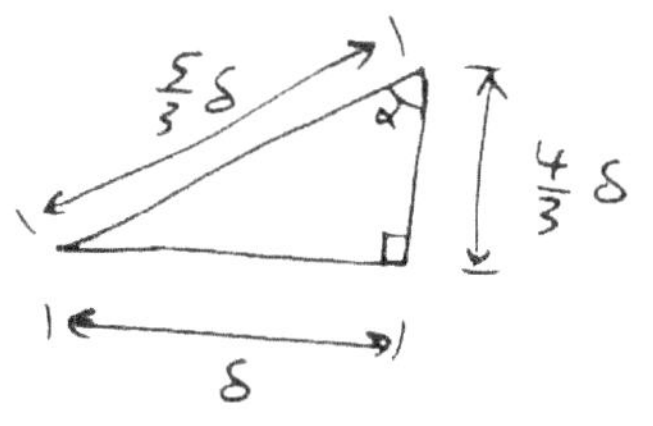

$\sin \alpha = \dfrac{15}{25} = \dfrac{3}{5}$

$\cos \alpha = \dfrac{20}{25} = \dfrac{4}{5}$

$\tan \alpha = \dfrac{3}{4}$

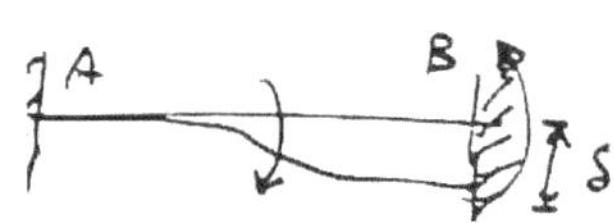

M_{AB} is negative

M_{BA} is negative

$M_{AB} = \dfrac{-6EI\delta}{L^2} = \dfrac{-6EI\,\delta}{(15)^2} = \dfrac{-6EI\delta}{225}$

$M_{BC} = \dfrac{+6EI}{L^2}\left(\dfrac{4\delta}{3}\right) = \dfrac{+6EI}{(20)^2}\left(\dfrac{4\delta}{3}\right) = \dfrac{+6EI\delta}{300}$

$M_{CD} = -\dfrac{6EI}{L^2}\left(\dfrac{5\delta}{3}\right) = \dfrac{-6EI}{(25)^2}\left(\dfrac{5\delta}{3}\right) = \dfrac{-6EI\delta}{375}$

Take $EI\delta = 1000$ (as an estimate):

then $\quad M_{AB} = -\dfrac{6000}{225} = -26.67 \quad$ k-ft.

$M_{BC} = +\dfrac{6000}{300} = \quad +20.0 \quad$ k-ft.

$M_{CD} = \dfrac{-6000}{375} = \quad -16.0 \quad$ k-ft.

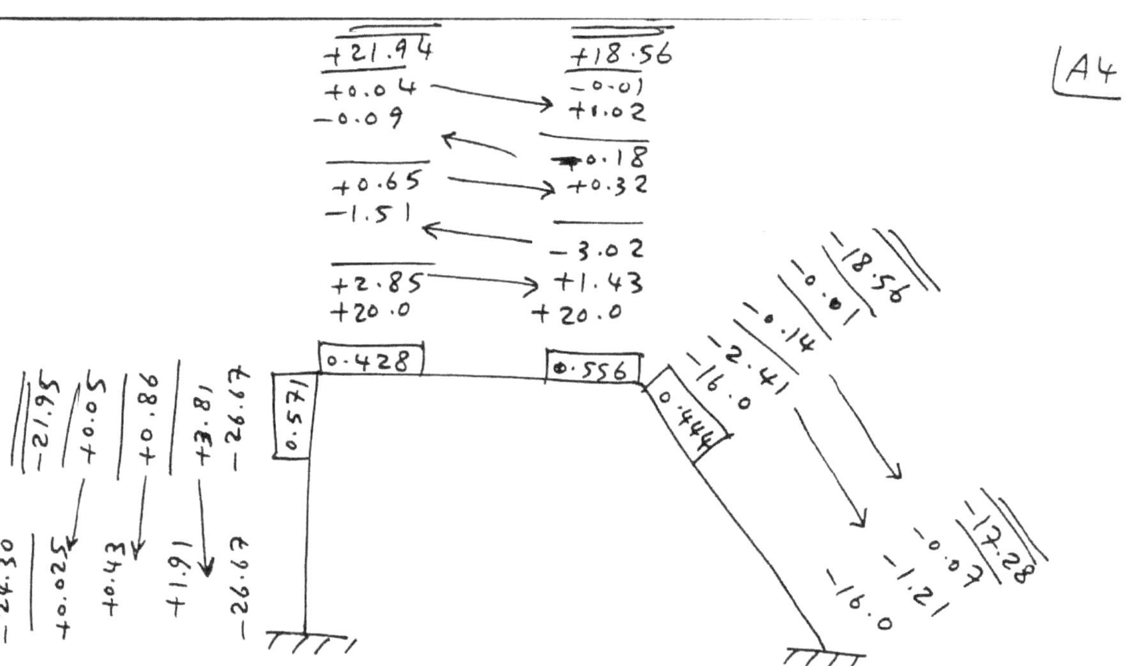

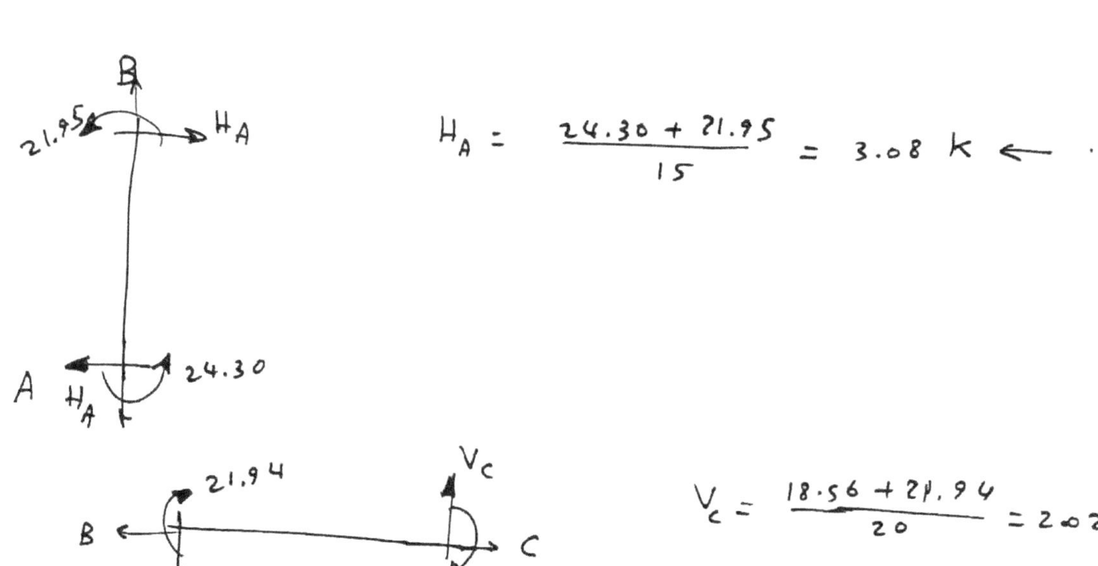

$$H_A = \frac{24.30 + 21.95}{15} = 3.08 \; k \; \leftarrow$$

$$V_C = \frac{18.56 + 21.94}{20} = 2.02 \; k$$

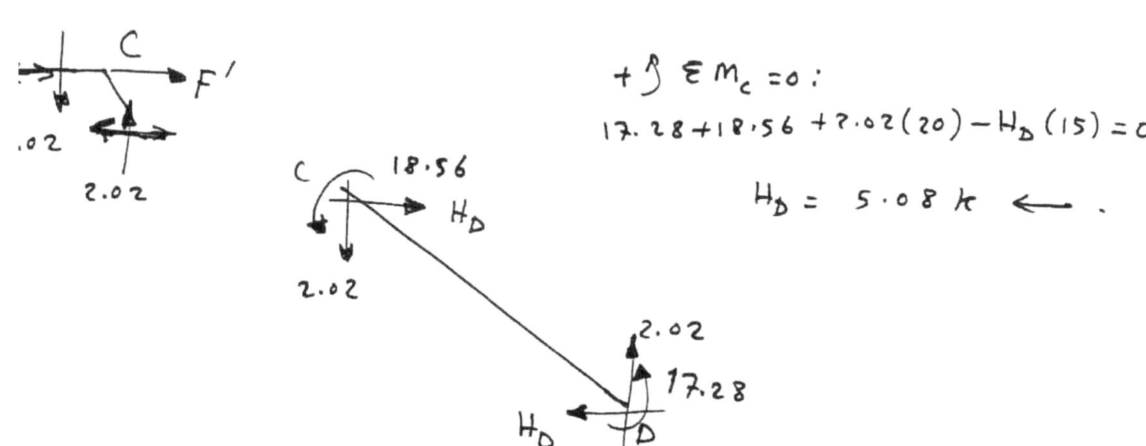

$$+\circlearrowleft \; \Sigma M_c = 0:$$

$$17.28 + 18.56 + 2.02(20) - H_D(15) = 0$$

$$H_D = 5.08 \; k \; \leftarrow$$

For the entire frame:

$$F' = 3.08 + 5.08 = 8.16 \text{ k}.$$

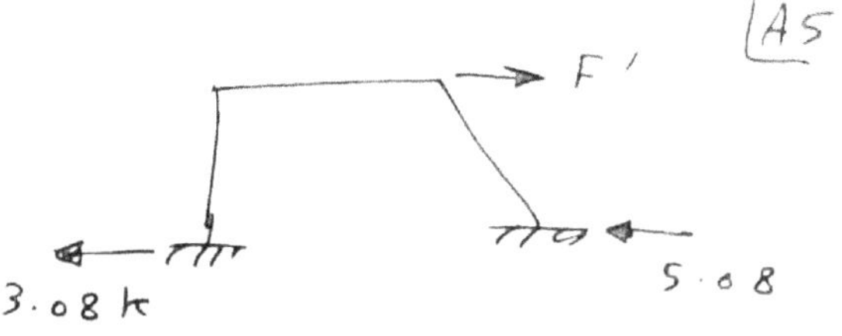

* Sidesway Correction Factor $\dfrac{F}{F'} = \dfrac{28.67}{8.16} = 3.51$.

$$\therefore \boxed{\text{Final Moments} = M_{step\ 1} + \left(\dfrac{F}{F'}\right) M_{step\ 2}}$$

$M_{AB} = -97.76 + (3.51)(-24.30) = -183.05 \text{ k-ft}.$
$\qquad \approx -183 \text{ k-ft}.$

$M_{BA} = +29.47 + (3.51)(-21.95) = -47.57 \text{ k-ft}.$
$\qquad \approx -47.6 \text{ k-ft}.$

$M_{BC} = -29.4 + (3.51)(21.94) = +47.61 \text{ k-ft}.$
$\qquad \approx +47.6 \text{ k-ft}.$

$M_{CB} = -7.58 + (3.51)(18.56) = +57.56 \text{ k-ft}.$

$M_{CD} = +7.58 + (3.51)(-18.56) = -57.56 \text{ k-ft}.$

$M_{DC} = +3.78 + (3.51)(-17.28) = -56.87 \text{ k-ft}.$

Support Settlement:

Example: Determine the joint moments for the continuous beam shown in the figure if supports B and C settle 30 mm and 50 mm, respectively. Let $EI = 8000$ kN·m² for all members.

Solution:

Joint B:

$K_{AB} = I/6$

$K_{BC} = 2I/8$

$\Sigma K_B = I/6 + 2I/8 = 0.417 I$

$DF_{AB} = \dfrac{I/6}{0.417 I} = 0.4$

$DF_{BC} = \dfrac{2I/8}{0.417 I} = 0.6$

Joint C:

$K_{CD} = I/8$

$\Sigma K_C = 2I/8 + I/8 = 3I/8$

$DF_{BC} = \dfrac{2I/8}{3I/8} = \dfrac{2}{3}$

$DF_{CD} = \dfrac{I/8}{3I/8} = \dfrac{1}{3}$

Fixed-End Moments due to Support Settlement:

$M^F_{AB} = M^F_{BA} = -6\dfrac{EI\delta}{L^2}$

$\quad = -\dfrac{6(8000)\left(\frac{30}{1000}\right)}{(6)^2}$

$\quad = -40$ kN·m.

$M^F_{BC} = M^F_{CB} = -6\dfrac{E(2I)\delta}{L^2}$

$\quad = -\dfrac{6(2)(8000)\left(\frac{20}{1000}\right)}{(8)^2}$

$\quad = -30$ kN·m.

$M^F_{CD} = M^F_{DC} = +6\dfrac{EI\delta}{L^2}$

$\quad = \dfrac{6(8000)(50/1000)}{(8)^2}$

$\quad = +37.5$ kN·m

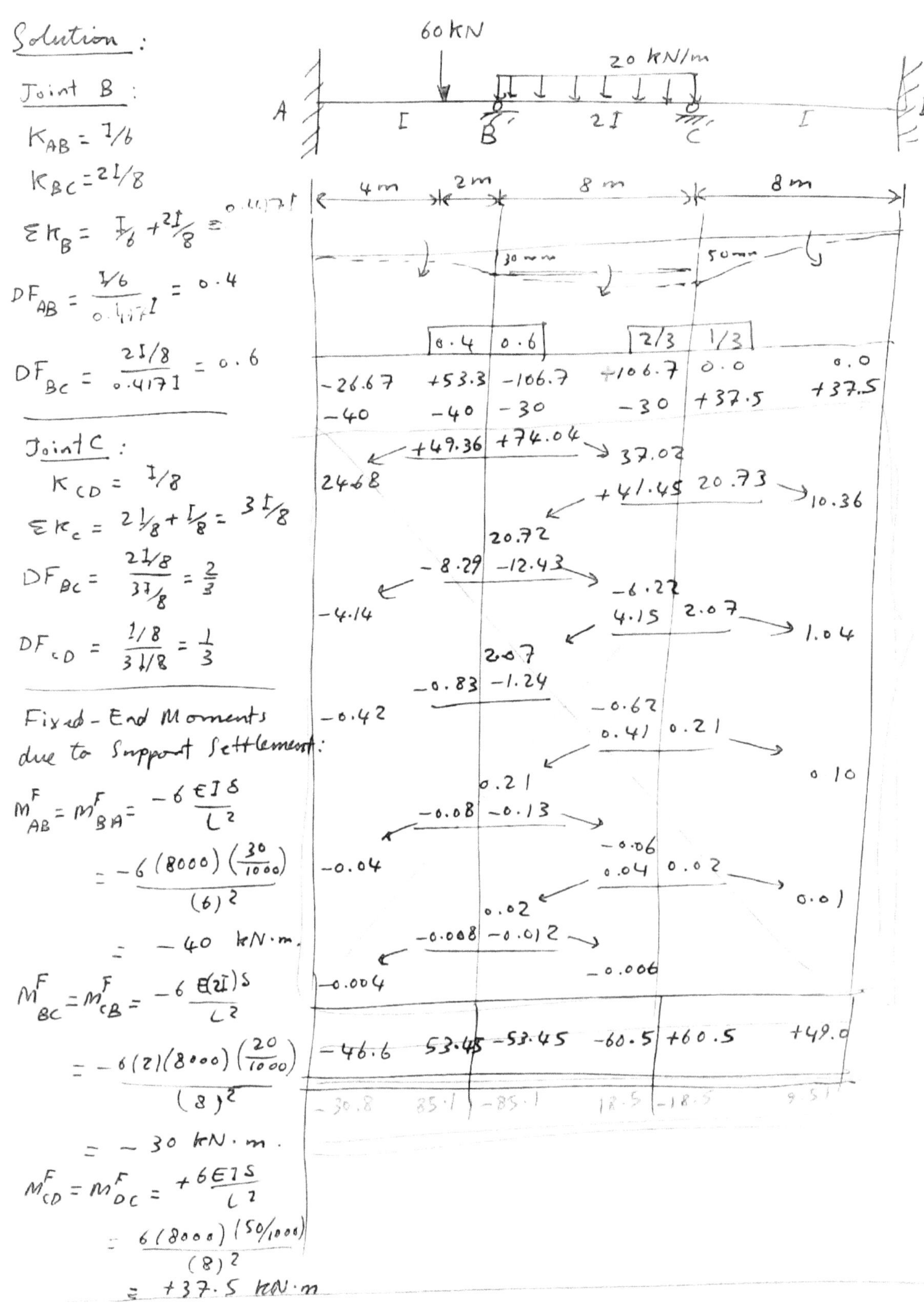

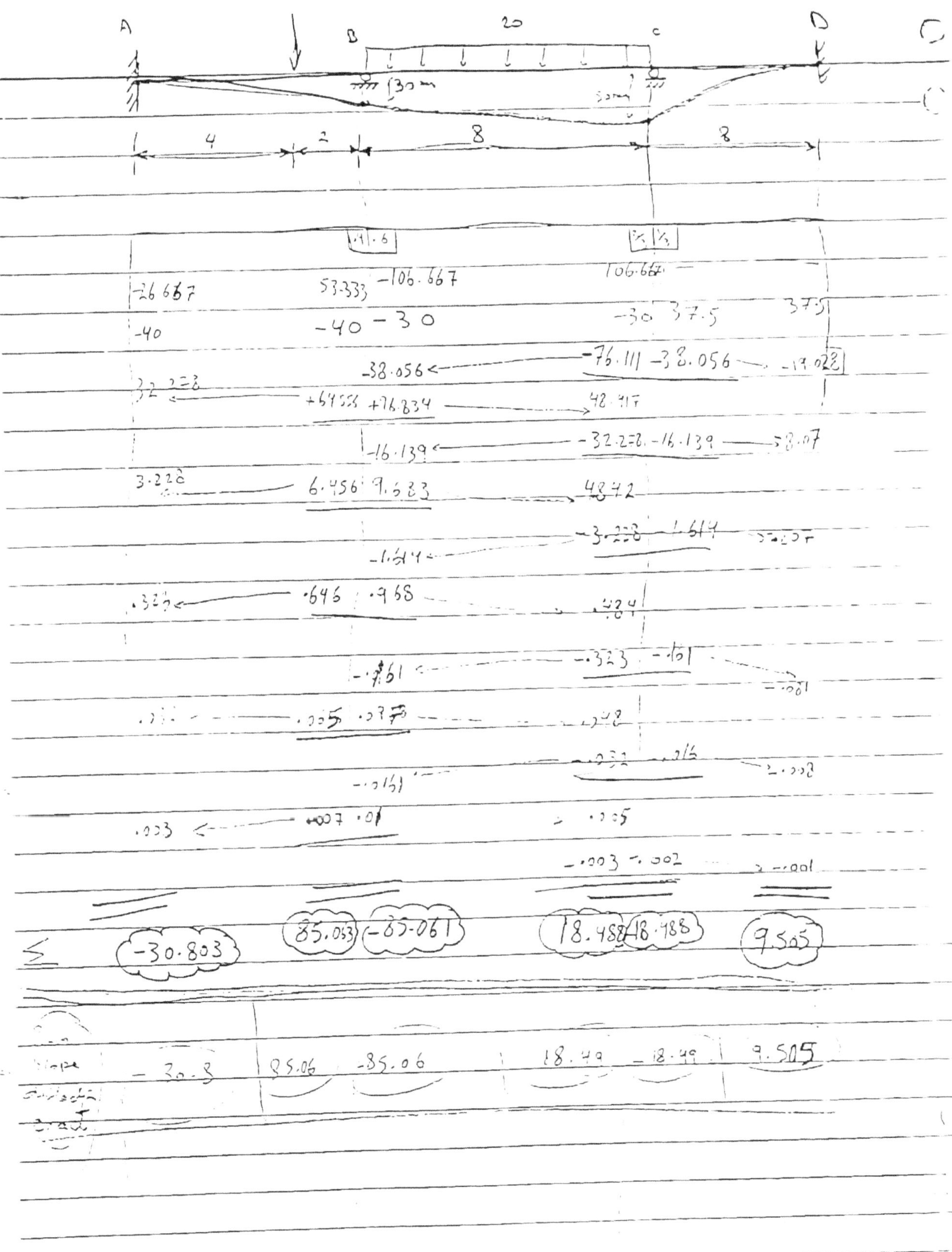

Multi-Storey Frames:

Example (method of Solution ONLY):

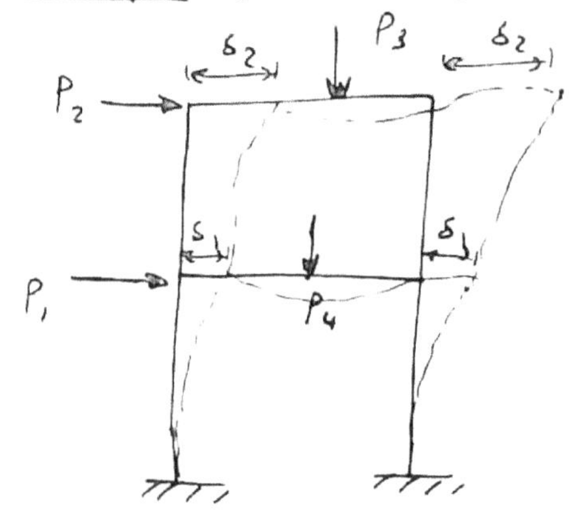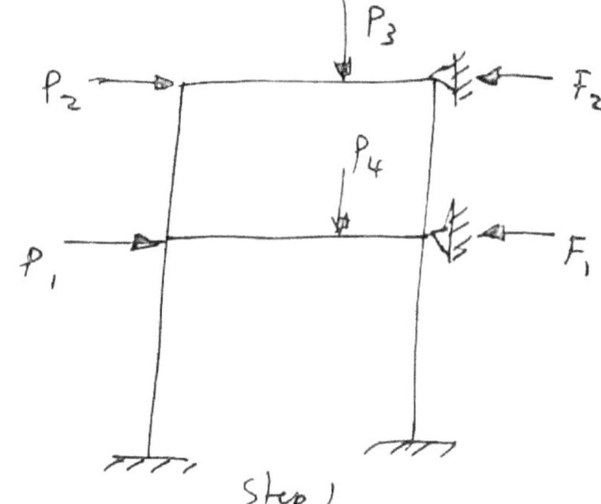

$=$

$+$ 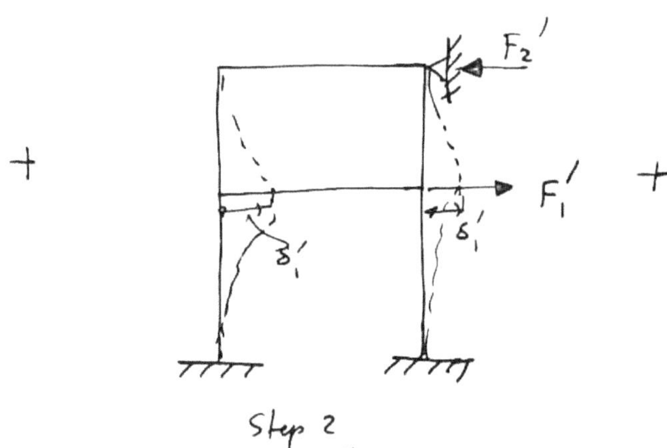 $+$ 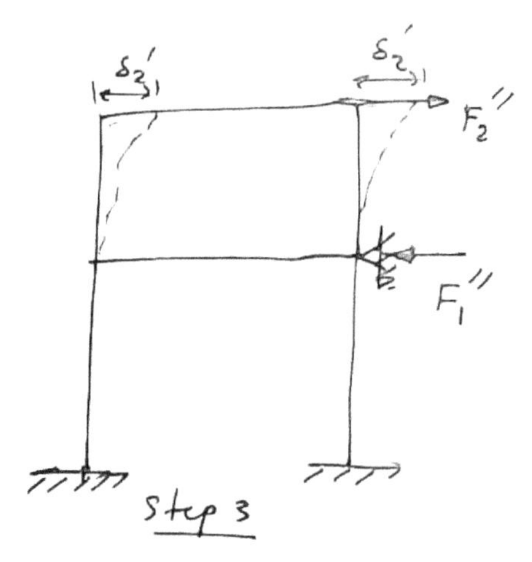

The above frame has **two degrees of sidesway**.

At each imposed restraint (at each storey):

At level (storey 1): $\quad -F_1 + k_1 F_1' - k_2 F_1'' = 0 \quad\text{——— (1)}$

At level (storey 2): $\quad -F_2 - k_1 F_2' + k_2 F_2'' = 0 \quad\text{——— (2)}$

Solve the above two equations simultaneously to obtain k_1 and k_2 (sidesway correction factors):

$$\boxed{\text{Final Moments} = M_{step\,1} + k_1 \cdot M_{step\,2} + k_2 \cdot M_{step\,3}}$$

Matrix structural computer analysis provides an efficient and economical alternative for the analysis of multistory frame buildings. An analyst can write a program or use one of the many available mainframe or mini-computer codes. The IMAGES-2D program represents one choice for multistory plane-frame structural analyses. Laursen [13.4] provides listings of two microcomputer BASIC and FORTRAN matrix programs for the solution of simultaneous equations and moment-distribution analysis. Lefter and Bergin [13.5] also present a BASIC program for the solution of simultaneous equations along with software to perform plane-frame analysis. Consequently, this author strongly recommends the use of a computer for the analysis of multistory frames with sidesway.

13.8. NONPRISMATIC BEAMS

Thus far, the method of moment distribution has been limited to prismatic structures. Yet, bridge and building structures often contain nonprismatic members which are identified by a variable depth along their span lengths (see Fig. 13.8). A variable member depth is usually selected to lower the stress at points of high bending moment and to maintain deflections within acceptable limits. Now, we will study the application of moment distribution for the analysis of nonprismatic indeterminate beam structures.

The basic concepts of moment distribution remain the same for both prismatic and nonprismatic members. However, the fixed-end moments,

Figure 13.8 A three-span continuous concrete highway bridge in Wilmore, Kentucky, with an increasing variable depth shown over the two interior supports. (*Courtesy of International Structural Slides, Berkeley.*)

Analysis of Nonprismatic Members by the Moment Distribution Method:

* Thus far, the method of moment distribution has been limited to prismatic structures. Yet, bridge and building structures often contain nonprismatic members which are identified by a variable depth along their span lengths.

* The basic concepts of moment distribution remain the same for both prismatic and nonprismatic members.

* Note: the fixed-end moments, distribution factors, and carry-over factors are not valid for nonprismatic members.

* Therefore, the fixed-end moments (FEMs), distribution factors and carry-over factors must first be determined for each nonprismatic member before a moment distribution analysis of the structure can be done.

___

Definitions:

(1) A carry-over factor is the ratio of the moment that develops at the fixed-end of a member to the applied moment at the simply-supported end.

C_{AB} : carry-over factor of member AB at end A.
C_{BA} : carry-over factor of member AB at end B.

(2) Absolute Stiffness : is the moment applied at the simple end of a pin-fixed member that will produce a unit rotation at the simply-supported end.

K_{AB} : stiffness factor of member AB at end A

$$\text{Absolute Stiffness} = K_{AB} \frac{E I_c}{L}$$

where $I_c \equiv$ moment of inertia at minimum depth.

K_{BA} : stiffness factor of member AB at end B. |A9

---

* For any prismatic member :

$$\text{Carry-Over Factor} = +\tfrac{1}{2}.$$

$$\text{Absolute Stiffness} = \frac{4EI}{L}.$$

Example 1 :

Determine carry-over factors, absolute stiffnesses, and FEMs at each end of the non-prismatic beams shown in the figure for a uniform load w.

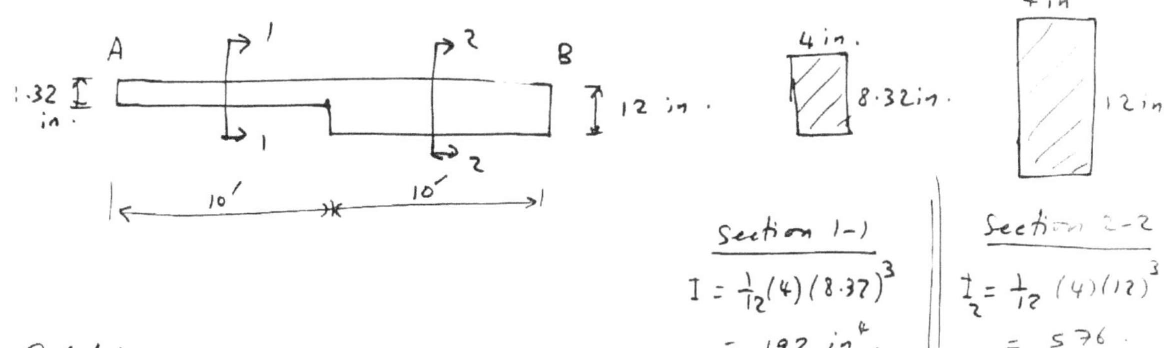

Solution :

Carry-Over Factors :

1. Carry-Over Factor C_{AB} :

$$\boxed{M_{BA} = C_{AB}\,M}$$

$$\Rightarrow \quad C_{AB} = \frac{M_{BA}}{M}.$$

Apply M at A, Find M_{BA} at B ?

Use the method of consistent deformations :

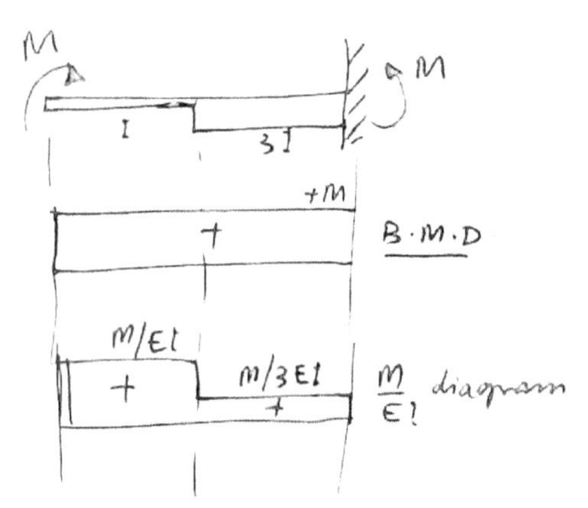

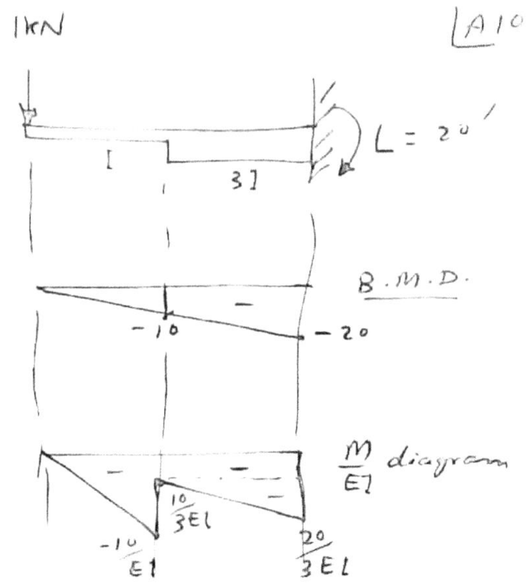

To calculate the required deflections, use the moment area theorems:

$$\delta_{A_0} = \left(\frac{M}{EI}\right)(10)(5) + \left(\frac{M}{3EI}\right)(10)(15) = +\frac{100M}{EI} \quad (up).$$

$$f_{AA} = -\frac{1}{2}\left(\frac{10}{EI}\right)(10)\left(\frac{2}{3}\right)(10) - \left(\frac{10}{3EI}\right)(10)(15) - \frac{1}{2}\left(\frac{10}{3EI}\right)(10)\left(10 + \frac{2}{3}(10)\right)$$

$$= -\frac{1111}{EI} \quad (down).$$

$$\Rightarrow \quad \delta_A = 0 = \delta_{A_0} + R_A f_{AA} = 0$$

$$\frac{100M}{EI} + R_A\left(-\frac{1111}{EI}\right) = 0$$

$$\Rightarrow \quad R_A = 0.090M \uparrow$$

For the entire (original) beam:

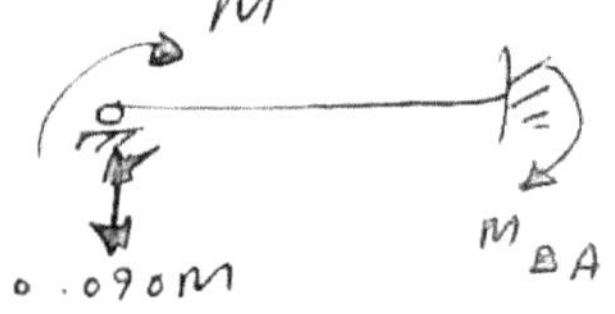

$$+^\curvearrowright \; \Sigma M_B = 0:$$

$$M_{BA} = -M + 0.090M(20)$$

$$M_{BA} = +0.80M = C_{AB}M$$

$$\Rightarrow \quad \boxed{C_{AB} = 0.80} \quad \checkmark$$

2 Carry-Over Factor C_{BA}:

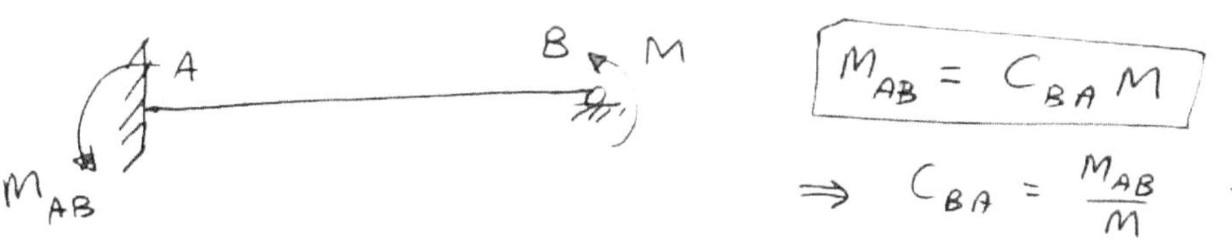

$$M_{AB} = C_{BA} M$$

$$\Rightarrow \quad C_{BA} = \frac{M_{AB}}{M}$$

Apply M at B, Find M_{AB} at A ?

Use the method of consistent deformations:

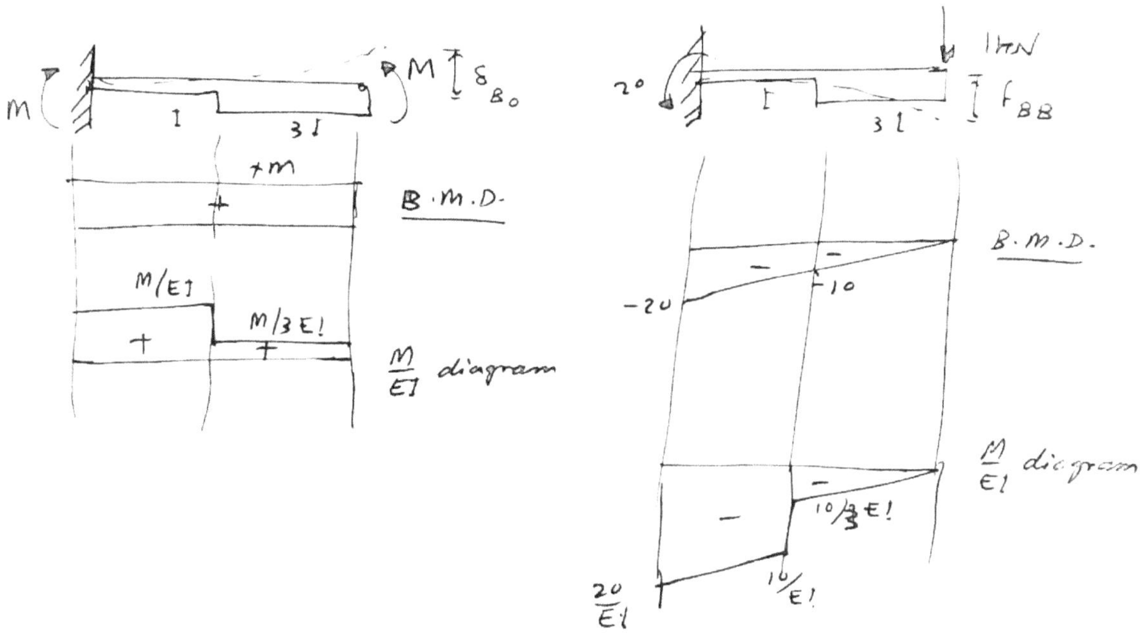

$$\delta_{B_0} = + \left(\frac{M}{3EI}\right)(10)(5) + \left(\frac{M}{EI}\right)(10)(15) = + \frac{166.67\,M}{EI} \quad (up)$$

$$f_{BB} = -\frac{1}{2}\left(\frac{10}{3EI}\right)(10)\left(\frac{2}{3}\right)(10) - \left(\frac{10}{3EI}\right)(10)(15) - \frac{1}{2}\left(\frac{10}{EI}\right)(10)\left(10 + \frac{2}{3}(10)\right)$$

$$= -\frac{2444}{EI}$$

$$\Rightarrow \quad \delta_B = \delta_{B_0} + R_B\, f_{BB} = 0$$

$$\frac{166.67\,M}{EI} + R_B\left(-\frac{2444}{EI}\right) = 0$$

$$\Rightarrow \quad R_B = 0.0682\,M$$

For the original beam:

$+\circlearrowright \Sigma m_A = 0:$

$M_{AB} = 0.0682M \,(20) - M$

$M_{AB} = +0.364\,M$

$\qquad = C_{BA}\,M$

$\Rightarrow \quad \boxed{C_{BA} = +0.364}$

M_{AB}

$0.0682\,M$

Absolute Stiffnesses:

③ Stiffness Factor k_{AB}:

By the first moment area theorem:

$$\theta_A - \cancel{\theta_B}^{\,0} = \left(\frac{M}{EI}\right)(10) + \left(\frac{M}{3EI}\right)(10)$$

$$\qquad - R_A\left[\frac{1}{2}\left(\frac{10}{EI}\right)(10) + \frac{1}{2}\left(\frac{10}{3EI}+\frac{20}{3EI}\right)(10)\right] \quad R_A$$

$$\text{but} \quad R_A = 0.09\,M$$

M/EI $\qquad M/3EI$

$\dfrac{-10}{3EI}$ $\qquad \dfrac{-20}{3EI}$

$\dfrac{-10}{EI}$

$$\Rightarrow \quad \theta_A = \frac{4.33\,M}{EI}$$

$$\Rightarrow \quad M = \frac{EI\,\theta_A}{4.33} = \frac{EI\,\theta_A}{4.33}\cdot\frac{L}{L}$$

$$\qquad = \frac{EI\,\theta_A}{4.33\,L}\,(20)$$

$$\qquad = 4.615\,\frac{EI}{L}\,\theta_A.$$

$$\text{but} \quad \frac{EI}{L}\,k_{AB} = M \quad \text{when } \theta_A = 1 \text{ rad.}$$

$$\Rightarrow \quad \boxed{k_{AB} = 4.615}$$

4 <u>Stiffness Factor K_{BA}:</u>

By the first moment-area theorem:

$$\theta_B - \theta_A = \left(\frac{M}{EI}\right)(10) + \frac{M}{3EI}(10)$$

$$- R_B\left[\frac{1}{2}\left(\frac{10}{3EI}\right)(10) + \frac{1}{2}\left(\frac{10}{EI} + \frac{20}{EI}\right)(10)\right]$$

but $R_B = 0.0682\,M$

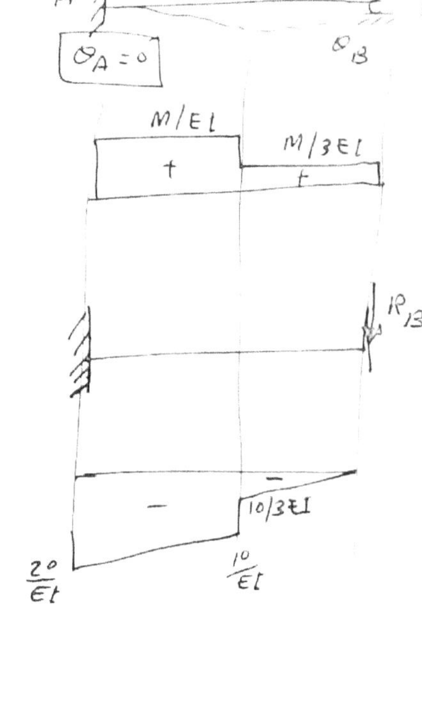

$$\Rightarrow \quad \theta_B = \frac{1.967\,M}{EI}$$

$$\therefore \quad M = \frac{EI\,\theta_B}{1.967} = \frac{EI\,\theta_B}{1.967}\cdot\frac{L}{L}$$

$$= \frac{EI\,\theta_B\,(20)}{1.967\,L}$$

$$= 10.17\,\frac{EI}{L}\,\theta_B$$

but $\quad M = K_{BA}\dfrac{EI}{L} \quad$ when $\theta_B = 1\,rad$

$$\Rightarrow \quad \boxed{K_{BA} = 10.17}$$

---

5 <u>Fixed-End Moments Due to Uniform Load w:</u>

Use the principle of superposition and the moment-area theorems:

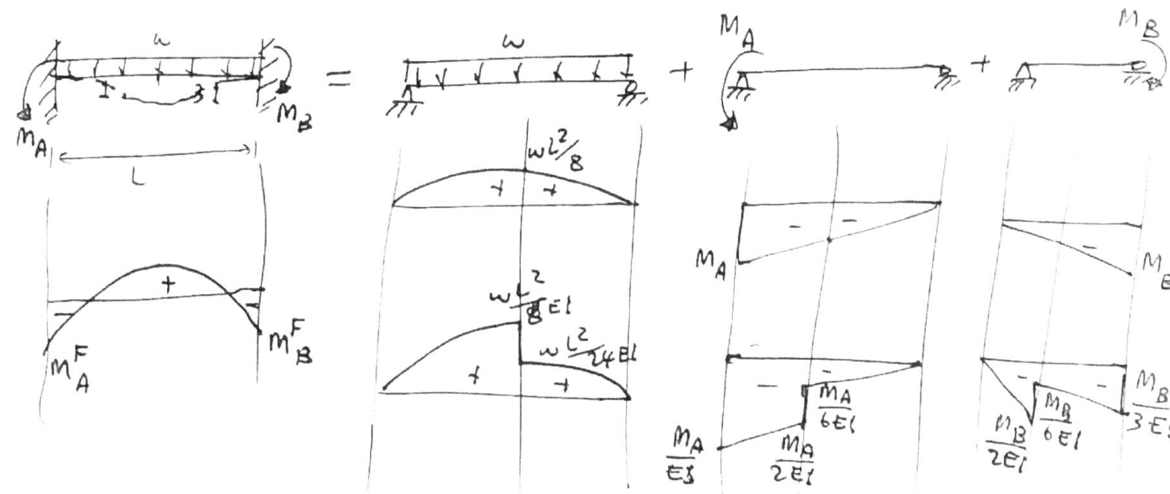

$$\theta_A - \theta_B = \frac{2}{3}\left(\frac{wL^2}{8EI}\right)(10) + \frac{2}{3}\left(\frac{wL^2}{24EI}\right)(10) - \frac{1}{2}\left(\frac{M_A}{EI} + \frac{M_A}{2EI}\right)(10)$$

$$- \frac{1}{2}\left(\frac{M_A}{6EI}\right)(10) - \frac{1}{2}\left(\frac{M_B}{2EI}\right)(10) - \frac{1}{2}\left(\frac{M_B}{6EI} + \frac{M_B}{3EI}\right)(10)$$

$$\Rightarrow \quad 0 = \frac{444.4w}{EI} - \frac{8.33 M_A}{EI} - \frac{5 M_B}{EI}$$

$$\Rightarrow \quad \boxed{2.33 M_A + 5 M_B = 444.4w} \quad \text{——— (1)}$$

$$S_{AB} = 0 = \frac{2}{3}\left(\frac{wL^2}{8EI}\right)(10)\left(\frac{5}{2}\right)(10) + \frac{2}{3}\left(\frac{wL^2}{24EI}\right)(10)\left(10 + \frac{3}{8}(10)\right)$$

$$- \left(\frac{M_A}{2EI}\right)(10)(5) - \frac{1}{2}\left(\frac{M_A}{2EI}\right)(10)\left(\frac{1}{3}\right)(10) - \frac{1}{2}\left(\frac{M_A}{6EI}\right)(10)\left(10 + \frac{1}{3}(10)\right)$$

$$- \frac{1}{2}\left(\frac{M_B}{2EI}\right)(10)\left(\frac{2}{3}\right)(10) - \left(\frac{M_B}{8EI}\right)(10)(15) - \frac{1}{2}\left(\frac{M_B}{6EI}\right)(10)\left(10 + \frac{2}{3}(10)\right)$$

$$\Rightarrow \quad 0 = \frac{3611.11w}{EI} - \frac{44.44 M_A}{EI} - \frac{55.56 M_B}{EI}$$

$$\Rightarrow \quad \boxed{44.44 M_A + 55.56 M_B = 3611.11w} \quad \text{——— (2)}$$

Solve equations (1) and (2) simultaneously for M_A and M_B

$$\begin{bmatrix} 2.33 & 5 \\ 44.44 & 55.56 \end{bmatrix} \begin{Bmatrix} M_A \\ M_B \end{Bmatrix} = \begin{Bmatrix} 444.4w \\ 3611.11w \end{Bmatrix}$$

Use program <u>SOLVE</u> to get:

$$\Rightarrow \quad M_A = 27.6w$$
$$M_B = 42.9w$$

<u>Note</u>: The final answer should be in terms of $\boxed{wL^2}$?

$$M_A = 27.6w \times \frac{L^2}{L^2} = \frac{27.6w L^2}{(20)^2} = 0.069 w L^2.$$

$$M_B = 42.9w \times \frac{L^2}{L^2} = \frac{42.9 w L^2}{(20)^2} = 0.107 w L^2.$$

<u>Note</u>. Compare the results with the FEMs for prismatic members.

$$m_A^F = m_B^F = \frac{1}{12} w L^2 = 0.083 \, w L^2.$$

<u>Example 2</u> :

Using the PCA tables, determine carry-over factors, stiffness factors, and FEMs at both ends of the nonprismatic beam shown in the figure. The load is uniformly distributed.

<u>Solution</u> :

Use <u>Table 13.3a</u> :

$$a_B = \frac{10}{20} = 0.5$$

$$r_B = \frac{12 - 8.32}{8.32} = 0.442$$

From the table :
Use <u>linear interpolation</u> :

$$C_{AB} = \frac{(0.442 - 0.4)(0.919 - 0.768)}{(0.6 - 0.4)}$$

$$+ 0.768$$

$$= 0.799$$
$$= 0.80$$

$$C_{BA} = \frac{(0.442 - 0.4)(0.343 - 0.371)}{(0.6 - 0.4)} + 0.371$$

$$= 0.365$$

$$k_{AB} = \frac{(0.442 - 0.4)(4.84 - 4.56)}{(0.6 - 0.4)} + 4.56$$

$$= 4.62$$

$$k_{BA} = \frac{(0.442 - 0.4)(10.94 - 9.45)}{(0.6 - 0.4)} + 9.45$$

$$= 10.18$$

Fixed-End Moments Due to w:

M_{AB} coefficient $= \dfrac{(0.442-0.4)(0.0651-0.0700)}{(0.6-0.4)} + 0.0700$

$= 0.069$

$\Rightarrow \quad M_{AB} = 0.069 w L^2$

M_{BA} coefficient $= \dfrac{(0.442-0.4)(0.1176-0.1042)}{(0.6-0.4)} + 0.1042$

$= 0.107$

$\Rightarrow \quad M_{BA} = 0.107 w L^2$

Example 3 :

Compute all joint moments by moment distribution using the PCA tables for the continuous bridge girder under uniform loading.

Solution :

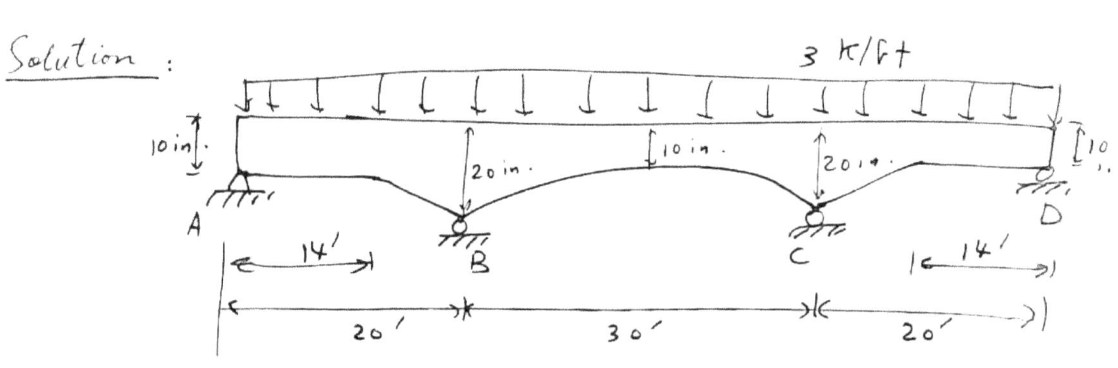

Member AB : $a_B = \dfrac{6}{20} = 0.3$

$a_A = 0$

$r_B = \dfrac{10}{10} = 1.0$ see [PCA Table 13.1]

$r_A = 0$

$\Rightarrow \quad C_{AB} = 0.791 \qquad k_{AB} = 4.71$

$\quad C_{BA} = 0.449 \qquad k_{BA} = 8.29$

$M_{AB} = 0.063 w L^2 = 0.063(3)(20)^2 = 75.6 \text{ k-ft.}$

$M_{BA} = 0.1311 w L^2 = 0.1311(3)(20)^2 = 157.3 \text{ k-ft.}$

Modified Stiffness $k'_{BA} = \dfrac{3}{4} k_{BA} = \dfrac{3}{4}(8.29) = 6.218$

$$K_{BA} = k_{BA}\frac{EI}{L} = \frac{6.218\,EI}{20} = 0.3109\,EI$$

Member BC : See PCA Table 13.2 :

$$a_A = \frac{15}{30} = 0.5 \qquad , \qquad a_B = \frac{15}{30} = 0.5$$

$$r_A = \frac{10}{10} = 1.0 \qquad , \qquad r_B = \frac{10}{10} = 1.0$$

$$\Rightarrow \quad C_{BC} = 0.694 \quad , \qquad C_{CB} = 0.694$$

$$k_{CB} = 12.03$$

$$k_{BC} = 12.03 \quad ,$$

$$M_{BC} = M_{CB} = 0.1025\,w\ell^2 = 0.1025\,(3)(30)^2 = 277\ \text{k-ft.}$$

$$K_{BC} = k_{BC}\frac{EI}{L} = \frac{12.03\,EI}{30} = 0.401\,EI$$

Joint B : $\sum k'_B = \;0.3109\,EI + 0.401\,EI = 0.7119\,EI$

$$DF_{AB} = \frac{0.3109\,EI}{0.7119\,EI} = \underline{\underline{0.437}}$$

$$DF_{BC} = \frac{0.401\,EI}{0.7119\,EI} = \underline{\underline{0.563}}$$

Due to symmetry, we have the same distribution factors
f joint C .

Continued $\Longrightarrow$

$\therefore$ By symmetry,

$$M_{BA} = M_{BC} = 260 \text{ k-ft.}$$

$$M_{CB} = M_{CD} = 260 \text{ k-ft.}$$

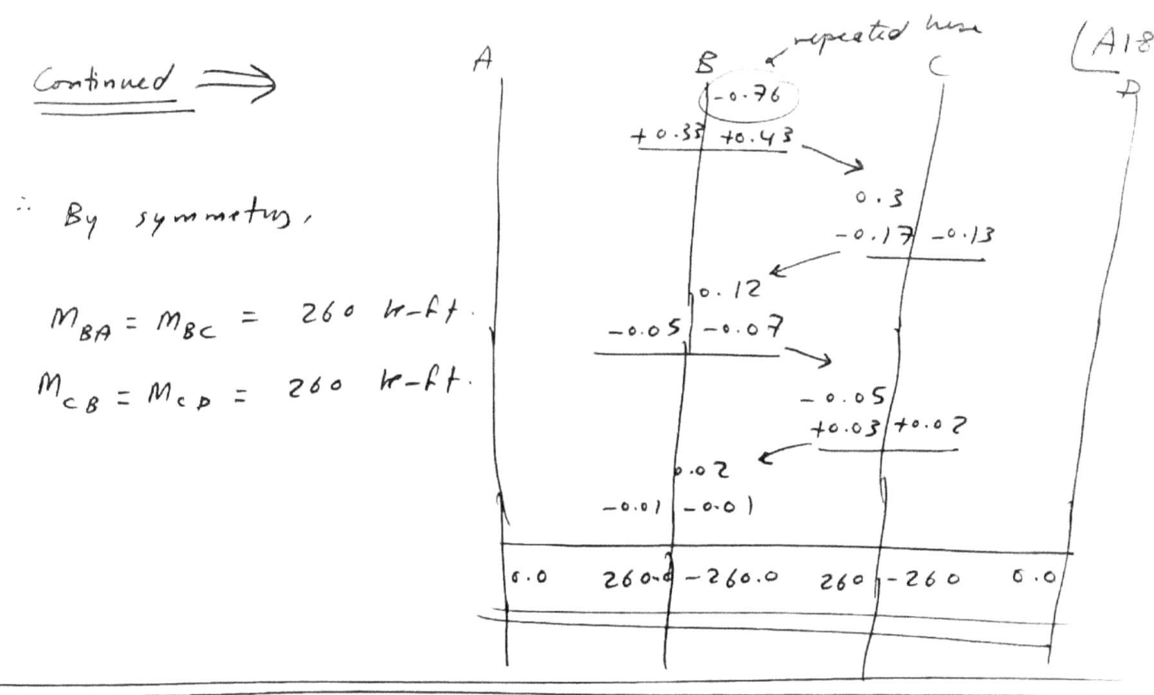

Chapter 15

Flexibility Matrix Method

Introduction:

* _Matrix analysis_ is a systematic approach to the problem of analyzing large and complex structures with the help of an electronic computer.

* Characteristics of matrix structural analysis.

  (1) all calculations in matrix analysis are carried out in _matrix algebra_.

  (2) the structure is _subdivided_ into _finite elements_.

* the flexibility matrix method is a _force_ method. (the forces are the unknowns in this method).

Flexibility Matrix:

$$\Delta_1 = F_{12} W_{12}$$

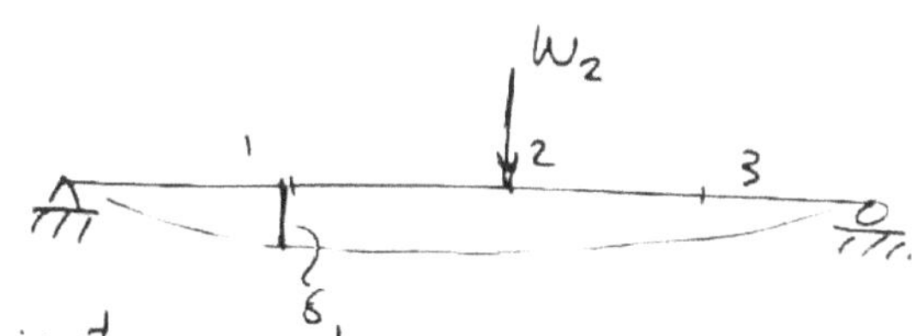

F_{12}: flexibility influence coefficient

F_{12} = deflection at 1 due to a unit load at 2.

$$\delta_1 = f_{11} W_1 + f_{12} W_2 + f_{13} W_3$$

$$\delta_2 = f_{21} W_1 + f_{22} W_2 + f_{23} W_3$$

$$\delta_3 = f_{31} W_1 + f_{32} W_2 + f_{33} W_3$$

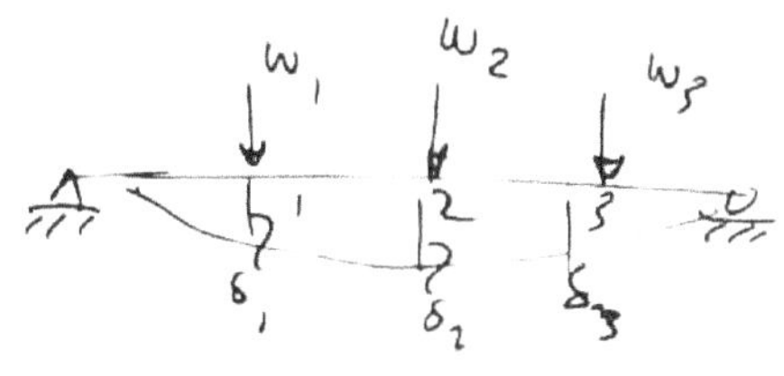

$$\Rightarrow \quad \begin{Bmatrix} \delta_1 \\ \delta_2 \\ \delta_3 \end{Bmatrix} = \begin{bmatrix} f_{11} & f_{12} & f_{13} \\ f_{21} & f_{22} & f_{23} \\ f_{31} & f_{32} & f_{33} \end{bmatrix} \begin{Bmatrix} W_1 \\ W_2 \\ W_3 \end{Bmatrix}$$

$$\Rightarrow \quad \boxed{\{\delta\} = [F]\{W\}}$$

$[F]$ = flexibility matrix

Maxwell's law of Reciprocal Deflections

$$\implies \quad f_{ij} = f_{ji}$$

$$\implies \text{the flexibility matrix } [f] \text{ is symmetric.}$$

<u>Example</u> :

Construct the flexibility matrix
for the beam shown in the figure.

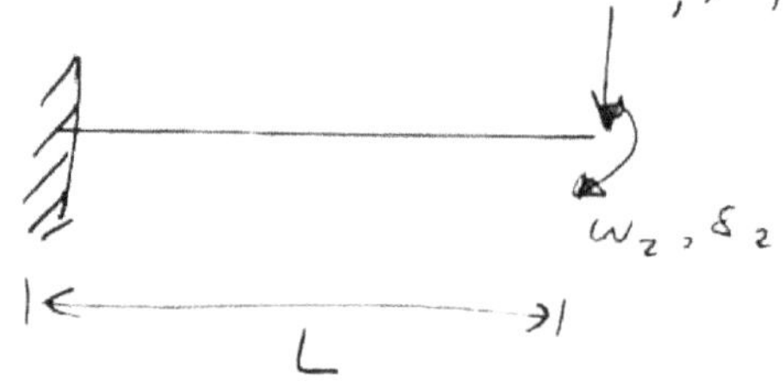

<u>Solution</u> :

W_1 , W_2 : generalized forces (forces & moments).

δ_1 , δ_2 : generalized deflections (deflections & rotations).

$$\begin{Bmatrix} \delta_1 \\ \delta_2 \end{Bmatrix} = \begin{bmatrix} f_{11} & f_{12} \\ f_{21} & f_{22} \end{bmatrix} \begin{Bmatrix} W_1 \\ W_2 \end{Bmatrix}$$

① Apply a unit vertical load at the free end.

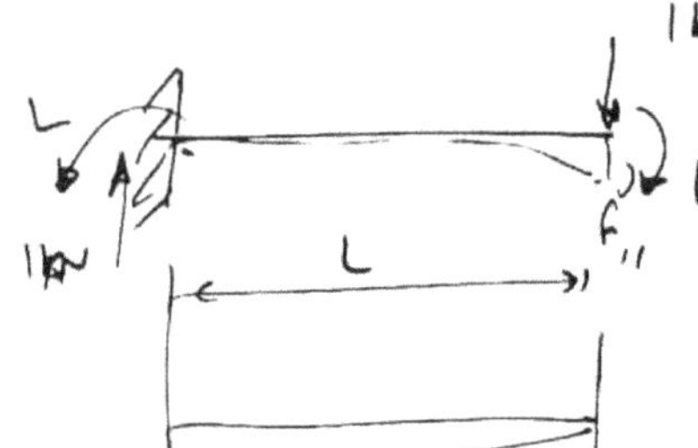

calculate f_{11} and f_{21} :

Use the moment-area theorems:

$$f_{21} = \frac{1}{2}\left(\frac{L}{EI}\right)(L) = \boxed{\frac{L^2}{2EI}}$$

$$f_{11} = \frac{1}{2}\left(\frac{L}{EI}\right)(L) \cdot \frac{2}{3}L = \boxed{\frac{L^3}{3EI}}$$

thus, we obtained the <u>first column</u>
of the <u>flexibility matrix</u>.

② Apply a unit moment at the free end :
calculate f_{12} and f_{22}.

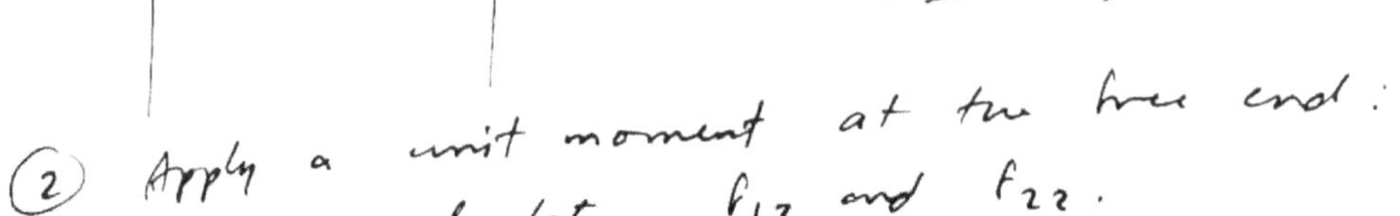

$$f_{22} = \frac{1}{EI}(L) = \boxed{\frac{L}{EI}}$$

$$f_{12} = \frac{1}{EI}(L)\left(\frac{L}{2}\right) = \boxed{\frac{L^2}{2EI}}$$

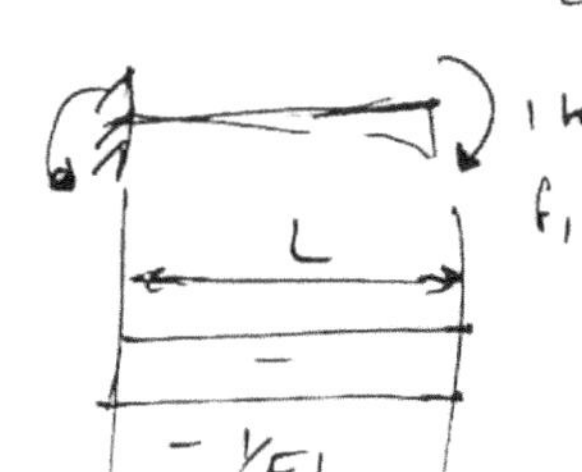

the flexibility matrix $[f]$ is given by:

$$[f] = \begin{bmatrix} \dfrac{L^3}{3EI} & \dfrac{L^2}{2EI} \\[2mm] \dfrac{L^2}{2EI} & \dfrac{L}{EI} \end{bmatrix}$$

and

$$\begin{Bmatrix} \delta_1 \\ \delta_2 \end{Bmatrix} = \begin{bmatrix} L^3/3EI & L^2/2EI \\[2mm] L^2/2EI & L/EI \end{bmatrix} \begin{Bmatrix} W_1 \\ W_2 \end{Bmatrix}$$

Example 15.1 :

Construct the flexibility matrix for the beam shown in the figure.

Solution :

* Generate the flexibility matrix column by column.

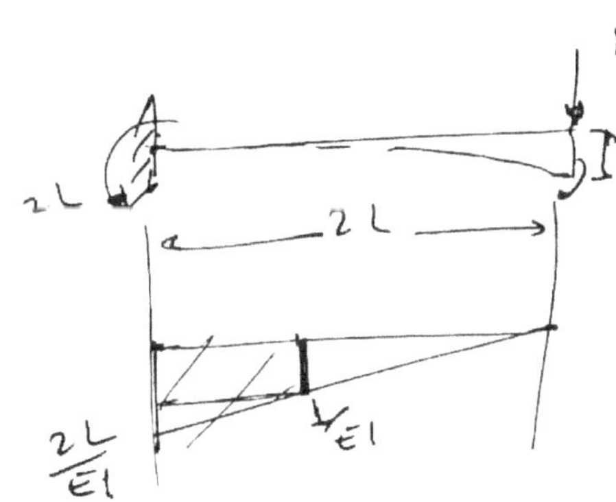

① First Column : f_{11} , f_{21} , f_{31} , f_{41} :

$$f_{11} = \frac{1}{2}\left(\frac{2L}{EI}\right)(2L)\left(\frac{2}{3}\,|\,(2L)\right) = \frac{8L^3}{3EI}$$

$$f_{21} = \frac{1}{2}\left(\frac{2L}{EI}\right)(2L) = \frac{2L^2}{EI}$$

$$f_{31} = \left(\frac{L}{EI}\right)(L)\left(\frac{L}{2}\right) + \frac{1}{2}\left(\frac{L}{EI}\right)(L)\left(\frac{2}{3}L\right)$$

$$= \frac{L^3}{2EI} + \frac{L^3}{3EI} = \frac{5L^3}{6EI}$$

$$f_{41} = \frac{1}{2}\left(\frac{L}{EI} + \frac{2L}{EI}\right)(L) = \frac{3L^2}{2EI}$$

$$f_{12} , f_{22} , f_{32} , f_{42}$$

② Second Column :

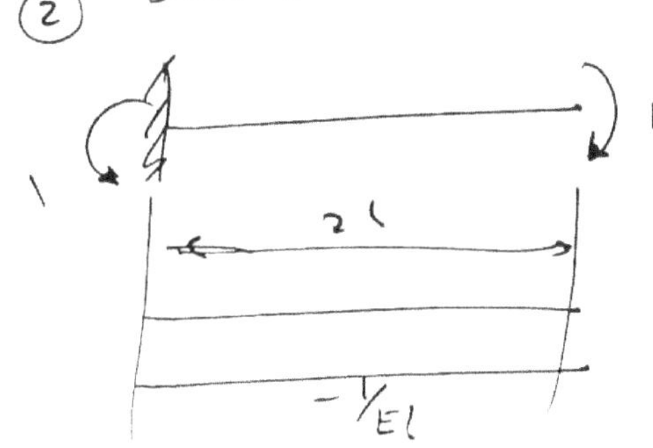

$$f_{12} = \left(\frac{1}{EI}\right)(2L)(L) = \frac{2L^2}{EI}$$

$$f_{22} = \left(\frac{1}{EI}\right)(2L) = \frac{2L}{EI}$$

$$f_{32} = \left(\frac{1}{EI}\right)(L)\left(\frac{L}{2}\right) = \frac{L^2}{2EI}$$

$$f_{42} = \left(\frac{1}{EI}\right)(L) = \frac{L}{EI}$$

③ Third Column: $f_{13}, f_{23}, f_{33}, f_{43}$:

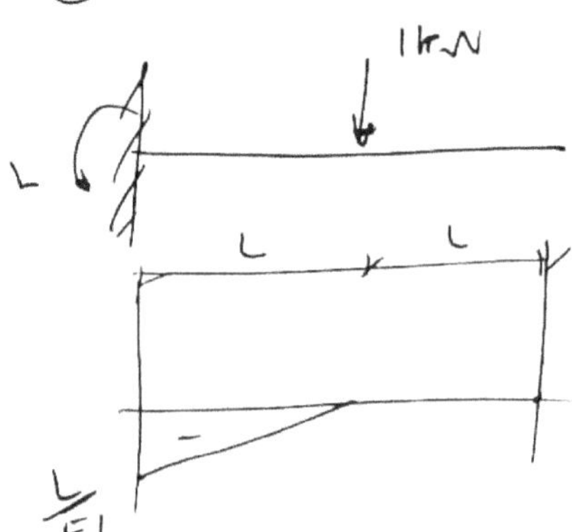

$$f_{13} = \frac{1}{2}\left(\frac{L}{EI}\right)(L)\left(L + \frac{2}{3}L\right)$$
$$= \frac{5L^3}{6EI}$$

$$f_{23} = \frac{1}{2}\left(\frac{L}{EI}\right)(L) = \frac{L^2}{2EI}$$

$$f_{33} = \frac{1}{2}\left(\frac{L}{EI}\right)(L)\left(\frac{2}{3}L\right) = \frac{L^3}{3EI}$$

$$f_{43} = \frac{1}{2}(L)\left(\frac{L}{EI}\right) = \frac{L^2}{2EI}$$

④ Fourth Column: $f_{14}, f_{24}, f_{34}, f_{44}$:

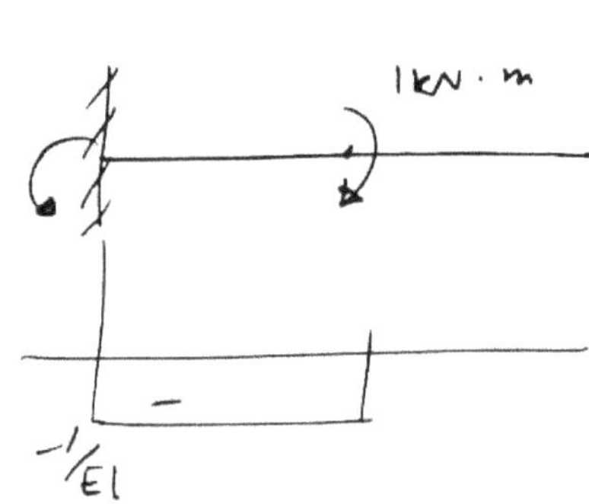

$$f_{14} = \left(\frac{1}{EI}\right)(L)\left(L + \frac{1}{2}L\right) = \frac{3L^2}{2EI}$$

$$f_{24} = \left(\frac{1}{EI}\right)(L) = \frac{L}{EI}$$

$$f_{34} = \left(\frac{1}{EI}\right)(L)\left(\frac{L}{2}\right) = \frac{L^2}{2EI}$$

$$f_{44} = \left(\frac{1}{EI}\right)(L) = \frac{L}{EI}$$

thus, the __structure-flexibility__ matrix is given by:

$$
\begin{Bmatrix} \delta_1 \\ \delta_2 \\ \delta_3 \\ \delta_4 \end{Bmatrix}
= \frac{1}{EI}
\begin{bmatrix}
\frac{8L^3}{3} & 2L^2 & \frac{5L^3}{6} & \frac{3L^2}{2} \\
2L^2 & 2L & \frac{L^2}{2} & L \\
\frac{5L^3}{6} & \frac{L^2}{2} & \frac{L^3}{3} & \frac{L^2}{2} \\
\frac{3L^2}{2} & L & \frac{L^2}{2} & L
\end{bmatrix}
\begin{Bmatrix} W_1 \\ W_2 \\ W_3 \\ W_4 \end{Bmatrix}
$$

Formation of the Structure-Flexibility Matrix from Element-Flexibility Matrices:

* there is difficulty in obtaining the required flexibility coefficients (deflections) for large and complex structures.

* the structure is <u>subdivided</u> into several <u>elements</u>, and a flexibility matrix is formed for each of the elements.

* the flexibility matrix for the entire structure is then obtained by combining the flexibility matrices of the individual elements.

Element and Structure Forces and Deformations:

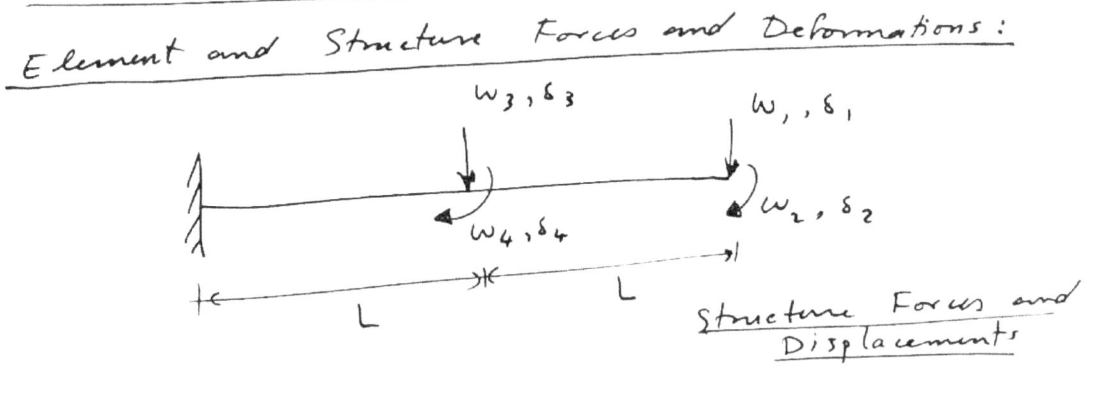

Obtain $[f]$ such that $\{\delta\} = [f]\{w\}$

$\uparrow$ "structure-flexibility matrix"

Subdivide the beam into two elements:

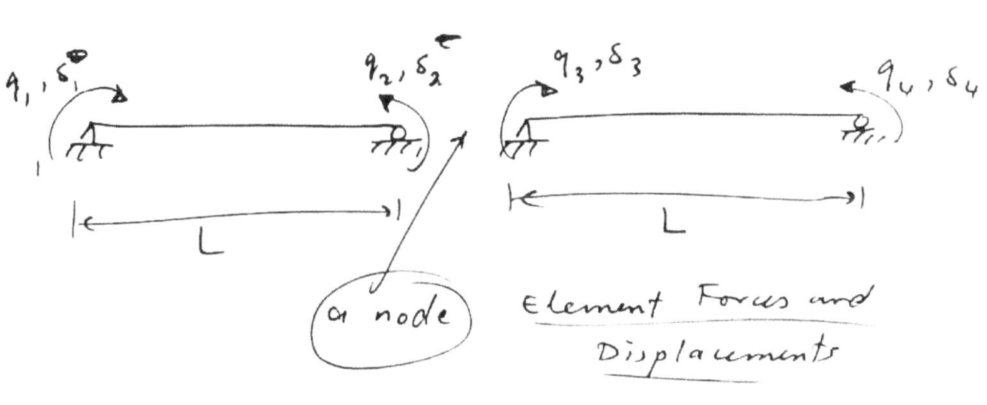

* Axial deformations are neglected.

* The junction of two elements is called a __node__.

* Assume all structure loads to be applied only at nodes.

Element — Flexibility Matrix:

Generate the element-flexibility matrix $[f_{ij}]$ for each element, column by column.

First Column:

$$f_{ii} = \frac{d_1}{L}$$

$$d_1 = \frac{1}{2}\left(\frac{1}{EI}\right)(L)\left(\frac{2}{3}L\right) = \frac{L^2}{3EI}$$

$$\Rightarrow f_{ii} = \boxed{\frac{L}{3EI}}$$

$$f_{ji} = \frac{d_2}{L}$$

$$d_2 = \frac{1}{2}\left(\frac{1}{EI}\right)(L)\left(\frac{1}{3}L\right) = \frac{L^2}{6EI}$$

$$\Rightarrow f_{ji} = \boxed{\frac{L}{6EI}}$$

Second Column:

$$f_{ji} = \frac{L}{6EI}$$

$$f_{jj} = \frac{L}{3EI}$$

$$[F] = \begin{bmatrix} f_{ii} & f_{ij} \\ f_{ji} & f_{jj} \end{bmatrix} = \begin{bmatrix} \dfrac{L}{3EI} & \dfrac{L}{6EI} \\ \dfrac{L}{6EI} & \dfrac{L}{3EI} \end{bmatrix} = \frac{L}{6EI}\begin{bmatrix} 2 & 1 \\ 1 & 2 \end{bmatrix}$$

$$\left\{\begin{array}{c} \delta_i \\ \delta_j \end{array}\right\} = \underbrace{\frac{1}{6EI}\begin{bmatrix} 2 & 1 \\ 1 & 2 \end{bmatrix}}_{f}\left\{\begin{array}{c} q_1 \\ q_2 \end{array}\right\}$$

"element flexibility matrix"

For each of the two elements we have:

For element ①:
$$\begin{Bmatrix} \delta_1 \\ \delta_2 \end{Bmatrix} = \frac{L}{6EI} \begin{bmatrix} 2 & 1 \\ 1 & 2 \end{bmatrix} \begin{Bmatrix} q_1 \\ q_2 \end{Bmatrix}$$

For element ②:
$$\begin{Bmatrix} q_3 \\ q_4 \end{Bmatrix} = \frac{L}{6EI} \begin{bmatrix} 2 & 1 \\ 1 & 2 \end{bmatrix} \begin{Bmatrix} q_3 \\ q_4 \end{Bmatrix}$$

Form the "composite element–flexibility matrix" as follows:

$$\begin{Bmatrix} \delta_1 \\ \delta_2 \\ ---- \\ \delta_3 \\ \delta_4 \end{Bmatrix} = \frac{L}{6EI} \begin{bmatrix} 2 & 1 & 0 & 0 \\ 1 & 2 & 0 & 0 \\ 0 & 0 & 2 & 1 \\ 0 & 0 & 1 & 2 \end{bmatrix} \begin{Bmatrix} q_1 \\ q_2 \\ --- \\ q_3 \\ q_4 \end{Bmatrix}$$

"block–diagonal matrix"

$$\boxed{\{\delta\} = [f_c]\{q\}}$$

Force – Transformation Matrix:

* the composite flexibility matrix $[f_c]$ describes the force–deformation characteristics of the individual elements, taken one at a time.

* Introduce the force–transformation matrix $[B]$, which relates what occurs in the elements to the behavior of the entire structure.

$$\begin{Bmatrix} q_1 \\ q_2 \\ q_3 \\ q_4 \end{Bmatrix} = \begin{bmatrix} B_{11} & B_{12} & B_{13} & B_{14} \\ B_{21} & B_{22} & B_{23} & B_{24} \\ B_{31} & B_{32} & B_{33} & B_{34} \\ B_{41} & B_{42} & B_{43} & B_{44} \end{bmatrix} \begin{Bmatrix} W_1 \\ W_2 \\ W_3 \\ W_4 \end{Bmatrix}$$

or $\{q\} = [B]\{w\}$

$[B] \equiv$ force-transformation matrix.

* Form $[B]$ one column at a time.

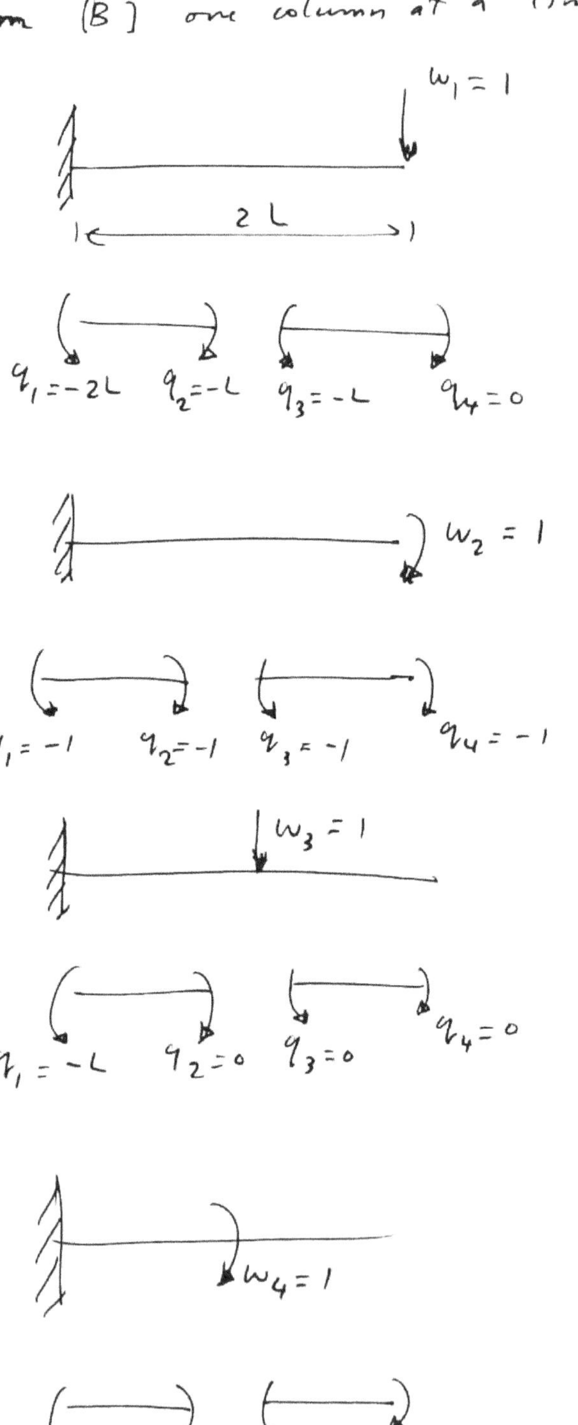

$$\Rightarrow \quad \begin{Bmatrix} q_1 \\ q_2 \\ q_3 \\ q_4 \end{Bmatrix} = \begin{bmatrix} -2L & -1 & -L & 1 \\ -L & -1 & 0 & 1 \\ -L & -1 & 0 & 0 \\ 0 & -1 & 0 & 0 \end{bmatrix} \begin{Bmatrix} w_1 \\ w_2 \\ w_3 \\ w_4 \end{Bmatrix}$$

<u>Transformation of $[f_e]$ into $[F]$:</u>

$\rightarrow$ Use the <u>principle of conservation of energy</u>:
Work performed by the structure forces acting on the entire structure $\equiv$ work of the element forces acting on all the elements into which the structure has been subdivided.

$$\frac{1}{2} [w_1 \ \ w_2 \ \ w_3 \ \ w_4] \begin{Bmatrix} \Delta_1 \\ \Delta_2 \\ \Delta_3 \\ \Delta_4 \end{Bmatrix} = \frac{1}{2} [q_1 \ \ q_2 \ \ q_3 \ \ q_4] \begin{Bmatrix} \delta_1 \\ \delta_2 \\ \delta_3 \\ \delta_4 \end{Bmatrix}$$

$$\Rightarrow \quad \frac{1}{2} \{w\}^T \{\Delta\} = \frac{1}{2} \{q\}^T \{\delta\}$$

but $\quad \{\delta\} = [f_e]\{q\}$

$$\{w\}^T \{\Delta\} = \{q\}^T [f_e]\{q\}$$

but $\quad \{q\} = [B]\{w\}$

$$\{w\}^T \{\Delta\} = \left([B]\{w\}\right)^T [f_e] [B]\{w\}$$

$$\{w\}^T \{\Delta\} = \{w\}^T [B]^T [f_e] [B]\{w\}$$

For arbitrary $\{w\}^T$:

$$\{\Delta\} = \left([B]^T [f_e] [B]\right)\{w\}$$
$$\overset{\downarrow}{[F]}$$
$$\{\Delta\} = [F]\{w\} \quad \text{where} \quad \boxed{[F] = [B]^T [f_e] [B]}$$

We have $[f_c] = \dfrac{L}{6EI}\begin{bmatrix} 2 & 1 & 0 & 0 \\ 1 & 2 & 0 & 0 \\ 0 & 0 & 2 & 1 \\ 0 & 0 & 1 & 2 \end{bmatrix}$

$[B] = \begin{bmatrix} -2L & -1 & -L & -1 \\ -L & -1 & 0 & -1 \\ -L & -1 & 0 & 0 \\ 0 & -1 & 0 & 0 \end{bmatrix}$

$$\boxed{[F] = [B]^T [f_c][B]}$$

$[f_c][B] = \dfrac{L}{6EI}\begin{bmatrix} 2 & 1 & 0 & 0 \\ 1 & 2 & 0 & 0 \\ 0 & 0 & 2 & 1 \\ 0 & 0 & 1 & 2 \end{bmatrix}\begin{bmatrix} -2L & -1 & -L & -1 \\ -L & -1 & 0 & -1 \\ -L & -1 & 0 & 0 \\ 0 & -1 & 0 & 0 \end{bmatrix}$

$= \dfrac{L}{6EI}\begin{bmatrix} -5L & -3 & -2L & -3 \\ -4L & -3 & -L & -3 \\ -2L & -3 & 0 & 0 \\ -L & -3 & 0 & 0 \end{bmatrix}$

$[F] = [B]^T[f_c][B] = \begin{bmatrix} -2L & -L & -L & 0 \\ -1 & -1 & -1 & -1 \\ -L & 0 & 0 & 0 \\ -1 & -1 & 0 & 0 \end{bmatrix}\dfrac{L}{6EI}\begin{bmatrix} -5L & -3 & -2L & -3 \\ -4L & -3 & -L & -3 \\ -2L & -3 & 0 & 0 \\ -L & -3 & 0 & 0 \end{bmatrix}$

$= \dfrac{L}{6EI}\begin{bmatrix} 16L^2 & 12L & 5L^2 & 9L \\ 12L & 12 & 3L & 6 \\ 5L^2 & 3L & 2L^2 & 3L \\ 9L & 6 & 3L & 6 \end{bmatrix}$

$\therefore \quad [F] = \dfrac{L}{EI}\begin{bmatrix} \dfrac{8L^3}{3} & 2L^2 & \dfrac{5L^3}{6} & \dfrac{3L^2}{2} \\[2mm] 2L^2 & 2L & \dfrac{L^3}{2} & L \\[2mm] \dfrac{5L^3}{6} & \dfrac{L^2}{2} & \dfrac{L^3}{3} & \dfrac{L^2}{2} \\[2mm] \dfrac{3L^2}{2} & L & \dfrac{L^2}{2} & L \end{bmatrix}$

Example 15.2 :

Construct the structure-flexibility matrix for the frame shown in the figure.

Solution :

Element loads & Displacements are given below.

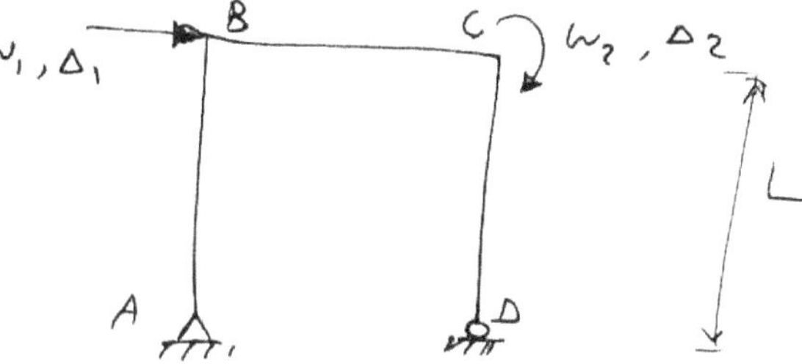

For element ①.
$$\begin{Bmatrix} \delta_1 \\ \delta_2 \end{Bmatrix} = \frac{L}{6EI} \begin{bmatrix} 2 & 1 \\ 1 & 2 \end{bmatrix} \begin{Bmatrix} q_1 \\ q_2 \end{Bmatrix}$$

For element ② :
$$\begin{Bmatrix} \delta_3 \\ \delta_4 \end{Bmatrix} = \frac{L}{6EI} \begin{bmatrix} 2 & 1 \\ 1 & 2 \end{bmatrix} \begin{Bmatrix} q_3 \\ q_4 \end{Bmatrix}$$

For element ③ :
$$\begin{Bmatrix} \delta_5 \\ \delta_6 \end{Bmatrix} = \frac{L}{6EI} \begin{bmatrix} 2 & 1 \\ 1 & 2 \end{bmatrix} \begin{Bmatrix} q_5 \\ q_6 \end{Bmatrix}$$

Composite Element Flexibility Matrix (f_c) is given by.

$$[f_c] = \frac{L}{6EI} \begin{bmatrix} 2 & 1 & & & & \\ 1 & 2 & & & & \\ & & 4 & 2 & & \\ & & 2 & 4 & & \\ & & & & 2 & 1 \\ & & & & 1 & 2 \end{bmatrix}$$

$$\begin{Bmatrix} \delta_1 \\ \delta_2 \\ \delta_3 \\ \delta_4 \end{Bmatrix} = \frac{L}{6EI} \begin{bmatrix} 2 & 1 & \varnothing \\ 1 & 2 & \\ \varnothing & 2 & 1 \\ & 1 & 2 \end{bmatrix} \begin{Bmatrix} q_1 \\ q_2 \\ q_3 \\ q_4 \end{Bmatrix}$$

$$\begin{Bmatrix} \delta_3 \\ \delta_4 \\ \delta_5 \\ \delta_6 \end{Bmatrix} = \frac{L}{6EI} \begin{bmatrix} 2 & 1 & \varnothing \\ 1 & 2 & \\ \varnothing & 2 & 1 \\ & 1 & 2 \end{bmatrix} \begin{Bmatrix} q_3 \\ q_4 \\ q_5 \\ q_6 \end{Bmatrix}$$

$$\begin{Bmatrix} \delta_1 \\ \delta_2 \\ \delta_3 \\ \delta_4 \\ \delta_5 \\ \delta_6 \end{Bmatrix} = \frac{L}{6EI} \begin{bmatrix} 2 & 1 & \varnothing & \\ 1 & 2 & & \\ & 2+2 & 1+1 & \\ \varnothing & 1+1 & 2+2 & \\ & & 2 & 1 \\ & & 1 & 2 \end{bmatrix} \begin{Bmatrix} q_1 \\ q_2 \\ q_3 \\ q_4 \\ q_5 \\ q_6 \end{Bmatrix}$$

$$\implies [f_c] = \frac{L}{6EI} \begin{bmatrix} 2 & 1 & \varnothing & \varnothing \\ 1 & 2 & & \\ \varnothing & 4 & 2 & \varnothing \\ & 2 & 4 & \\ \varnothing & \varnothing & 2 & 1 \\ & & 1 & 2 \end{bmatrix} \quad \text{Composite Element}$$
$$\text{Flexibility Matrix}$$

<u>Force - Transformation Matrix</u> : $[B]$ where $\{q\} = [B]\{w\}$

$$\{q\} = \begin{Bmatrix} q_1 \\ q_2 \\ q_3 \\ q_4 \\ q_5 \\ q_6 \end{Bmatrix} \qquad , \qquad \{w\} = \begin{Bmatrix} w_1 \\ w_2 \end{Bmatrix}$$

$$\underline{6 \times 1} \qquad \qquad \underline{2 \times 1}$$

$$\implies [B] \equiv \underline{6 \times 2} \text{ matrix.}$$

Generate $[B]$ column by column.

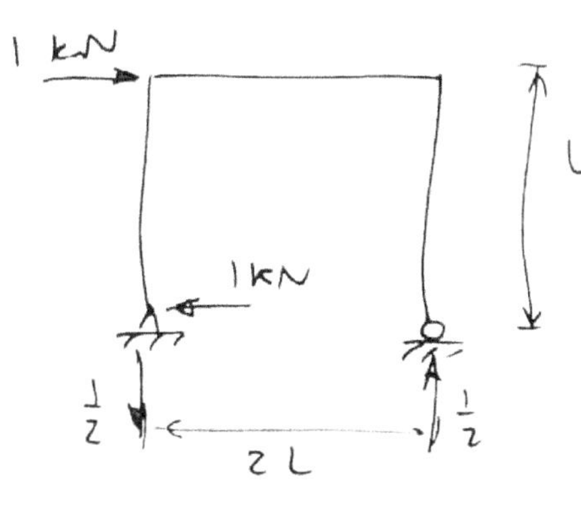

First column:

1 kN

1 kN

L

$\dfrac{1}{2}$ $\dfrac{1}{2}$

$2L$

$+L$

1

$\dfrac{1}{2}$

$$\left\{ \begin{array}{c} 0 \\ L \\ L \\ 0 \\ 0 \\ 0 \end{array} \right\}$$

$1\,kN\cdot m$

$\dfrac{1}{2L}$

$1\,kN\cdot m$

$\dfrac{1}{2L}$ $\dfrac{1}{2L}$

Second Column:

$$\left\{ \begin{array}{c} 0 \\ 0 \\ 0 \\ -1 \\ 0 \\ 0 \end{array} \right\}$$

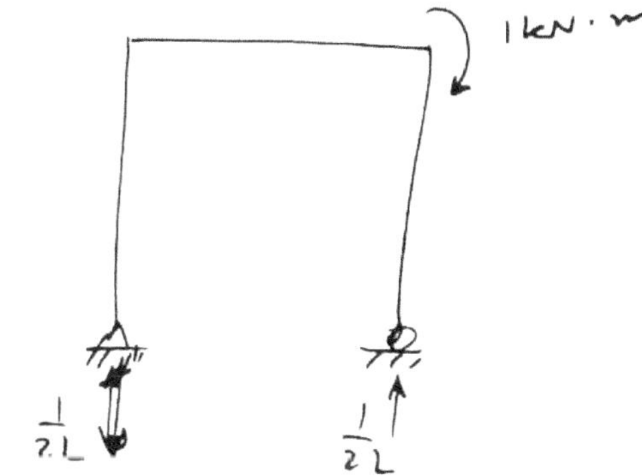

$$\Rightarrow \quad [B] = \left[\begin{array}{cc} 0 & 0 \\ L & 0 \\ L & 0 \\ 0 & -1 \\ 0 & 0 \\ 0 & 0 \end{array} \right]$$

Now use $\boxed{[F] = [B]^T \, [f_c] \, [B]}$

$$[f_c][B] = \frac{L}{6EI}\begin{bmatrix} 2 & 1 & 0 & 0 & 0 & 0 \\ 1 & 2 & 0 & 0 & 0 & 0 \\ 0 & 0 & 4 & 2 & 0 & 0 \\ 0 & 0 & 2 & 4 & 0 & 0 \\ 0 & 0 & 0 & 0 & 2 & 1 \\ 0 & 0 & 0 & 0 & 1 & 2 \end{bmatrix}_{6\times 6} \begin{bmatrix} 0 & 0 \\ L & 0 \\ L & 0 \\ 0 & -1 \\ 0 & 0 \\ 0 & 0 \end{bmatrix}_{6\times 2}$$

$$= \frac{L}{6EI}\begin{bmatrix} L & 0 \\ 2L & 0 \\ 4L & -2 \\ 2L & -4 \\ 0 & 0 \\ 0 & 0 \end{bmatrix}$$

$$[F] = [B]^T[f_c][B] = \begin{bmatrix} 0 & L & L & 0 & 0 & 0 \\ 0 & 0 & 0 & -1 & 0 & 0 \end{bmatrix}_{2\times 6} \cdot \frac{L}{6EI}\begin{bmatrix} L & 0 \\ 2L & 0 \\ 4L & -2 \\ 2L & -4 \\ 0 & 0 \\ 0 & 0 \end{bmatrix}_{6\times 2}$$

$$\therefore \ [F] = \frac{L}{6EI}\begin{bmatrix} 6L^2 & -2L \\ -2L & 4 \end{bmatrix}$$

$$\Rightarrow \ \begin{Bmatrix} \Delta_1 \\ \Delta_2 \end{Bmatrix} = \frac{L}{6EI}\begin{bmatrix} 6L^2 & -2L \\ -2L & 4 \end{bmatrix}\begin{Bmatrix} W_1 \\ W_2 \end{Bmatrix}$$

15.4 Analysis of Indeterminate Structures:

* the flexibility-matrix method is essentially a matrix formulation of the method of consistent deformations.

Example:

Choose M_A, M_C as the redundants.

Neglect N_A, N_C.

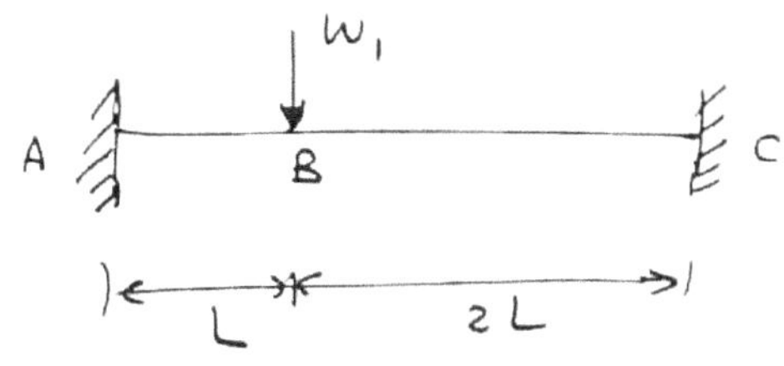

For element ①:

$$[f] = \frac{L}{6EI}\begin{bmatrix} 2 & 1 \\ 1 & 2 \end{bmatrix}$$

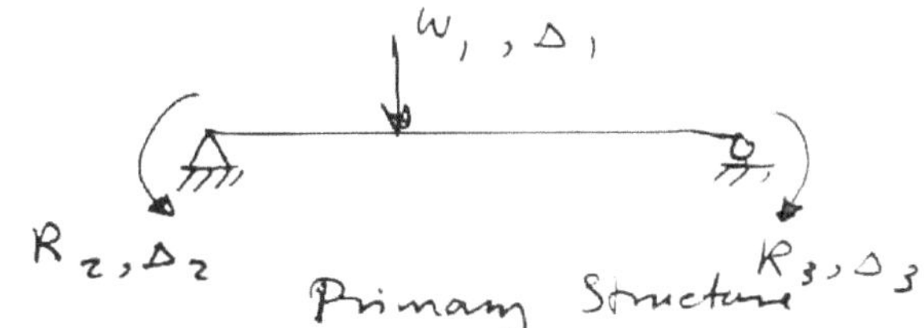

For element ②:

$$[f] = \frac{2L}{6EI}\begin{bmatrix} 2 & 1 \\ 1 & 2 \end{bmatrix}$$

$$= \frac{L}{6EI}\begin{bmatrix} 4 & 2 \\ 2 & 4 \end{bmatrix}$$

$$[f_c] = \frac{L}{6EI}\begin{bmatrix} 2 & 1 & 0 & 0 \\ 1 & 2 & 0 & 0 \\ 0 & 0 & 4 & 2 \\ 0 & 0 & 2 & 4 \end{bmatrix}$$

$$\{\Delta\} = [F]\{W\}$$

$$\begin{Bmatrix} \Delta_1 \\ \Delta_2 \\ \Delta_3 \end{Bmatrix} = \begin{bmatrix} F_{11} & F_{12} & F_{13} \\ F_{21} & F_{22} & F_{23} \\ F_{31} & F_{32} & F_{33} \end{bmatrix} \begin{Bmatrix} W_1 \\ R_2 \\ R_3 \end{Bmatrix}$$

Force-Transformation Matrix $[B]$:

$$\{q\} = [B]\{W\}$$

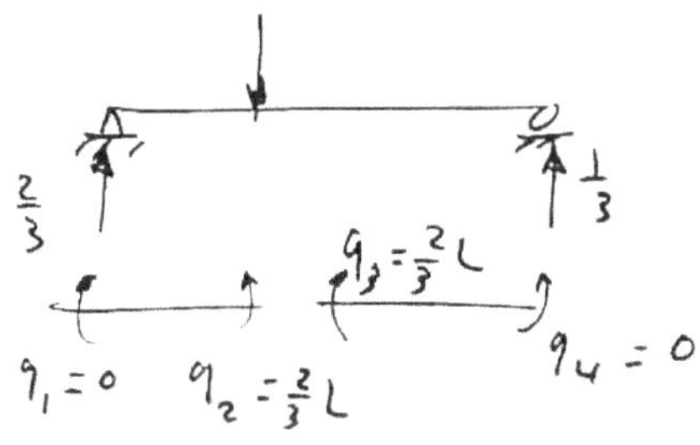

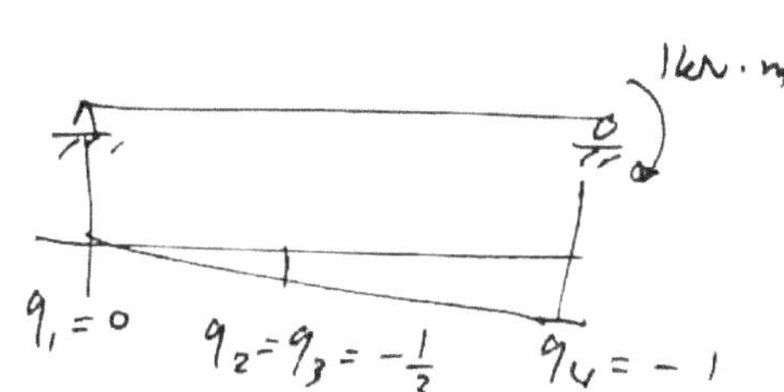

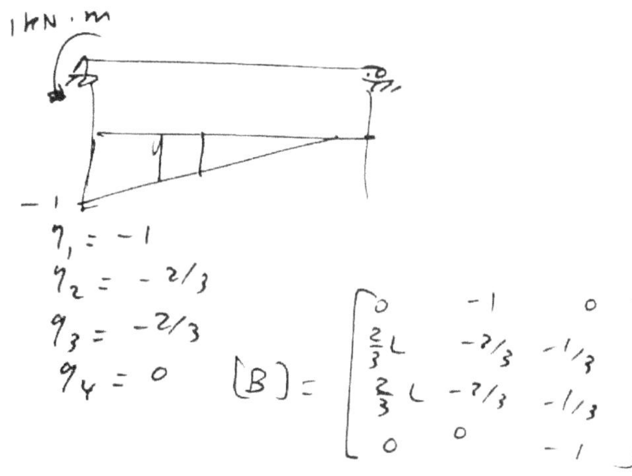

$$[B] = \begin{bmatrix} 0 & -1 & 0 \\ \frac{2}{3}L & -2/3 & -1/3 \\ \frac{2}{3}L & -2/3 & -1/3 \\ 0 & 0 & -1 \end{bmatrix}$$

$$[F] = [B]^T [f_c][B]$$

$$[f_c][B] = \frac{L}{6EI}\begin{bmatrix} 2 & 1 & 0 & 0 \\ 1 & 2 & 0 & 0 \\ 0 & 0 & 4 & 2 \\ 0 & 0 & 2 & 4 \end{bmatrix}\begin{pmatrix} 0 & 1 & 1 & 0 \\ 21/3 & -7/3 & -1/3 \\ 24/3 & -2/3 & -1/3 \\ 0 & 1 & 0 & -1 \end{pmatrix}$$

$$\underset{4\times 4}{} \qquad \underset{4\times 3}{}$$

$$= \frac{L}{6EI}\begin{bmatrix} 2\frac{L}{3} & -\frac{8}{3} & -\frac{1}{3} \\ 4L/3 & -7/3 & -\frac{2}{3} \\ 8L/3 & -8/3 & -\frac{10}{3} \\ 4L/3 & -4/3 & -14/3 \end{bmatrix}$$

$$[F] = \begin{bmatrix} 0 & 2\frac{L}{3} & 2\frac{L}{3} & 0 \\ -1 & -\frac{2}{3} & -\frac{2}{3} & 0 \\ 0 & -\frac{1}{3} & -\frac{1}{3} & -1 \end{bmatrix} \frac{L}{6EI}\cdot\begin{bmatrix} 2\frac{L}{3} & -\frac{8}{3} & -\frac{1}{3} \\ 4\frac{L}{3} & -7/3 & -\frac{2}{3} \\ 8\frac{L}{3} & -8/3 & -\frac{10}{3} \\ 4\frac{L}{3} & -4/3 & -\frac{14}{3} \end{bmatrix}$$

$$\underset{3\times 4}{} \qquad\qquad \underset{4\times 3}{}$$

$$= \frac{L}{6EI}\begin{bmatrix} \frac{24}{9}L^2 & -\frac{30L}{9} & -8L/3 \\ -\frac{30L}{9} & \frac{54}{9} & \frac{27}{9} \\ -\frac{8L}{3} & \frac{8}{3} & \frac{18}{3} \end{bmatrix}$$

$$\therefore [F] = \frac{L}{18EI}\begin{bmatrix} 8L^2 & -10L & -8L \\ -10L & 18 & 9 \\ -8L & 9 & 18 \end{bmatrix}$$

$$\Rightarrow \begin{Bmatrix} \Delta_1 \\ \hline D_2 \\ D_3 \end{Bmatrix} = \frac{L}{18EI}\begin{bmatrix} 8L^2 & -10L & -8L \\ -10L & 18 & 9 \\ -8L & 9 & 18 \end{bmatrix}\begin{Bmatrix} W_1 \\ \hline R_2 \\ R_3 \end{Bmatrix}$$

Partition the flexibility matrix into four submatrices.

Note: $\Delta_2 = 0$ and $\Delta_3 = 0$

$$\left\{ \begin{matrix} 0 \\ 0 \end{matrix} \right\} = \frac{L}{18EI} \left\{ \begin{matrix} -10L \\ -8L \end{matrix} \right\} [w_1] + \frac{L}{18EI} \begin{bmatrix} 18 & 9 \\ 9 & 18 \end{bmatrix} \left\{ \begin{matrix} R_2 \\ R_3 \end{matrix} \right\}$$

$$\Rightarrow \left\{ \begin{matrix} R_2 \\ R_3 \end{matrix} \right\} = - \begin{bmatrix} 18 & 9 \\ 9 & 18 \end{bmatrix}^{-1} \left\{ \begin{matrix} -10L \\ -8L \end{matrix} \right\} [w_1]$$

$$\quad\quad 2\times 2 \quad\quad 2\times 1 \quad\quad 1\times 2$$

but $\begin{bmatrix} 18 & 9 \\ 9 & 18 \end{bmatrix}^{-1} = \frac{1}{(18)^2-(9)^2} \begin{bmatrix} 18 & -9 \\ -9 & 18 \end{bmatrix}$

$$= \frac{1}{(18-9)(18+9)} \begin{bmatrix} 18 & -9 \\ -9 & 18 \end{bmatrix}$$

$$= \frac{1}{(9)(27)} \begin{bmatrix} 18 & -9 \\ -9 & 18 \end{bmatrix}$$

$$= \frac{9}{9(27)} \begin{bmatrix} 2 & -1 \\ -1 & 2 \end{bmatrix}$$

$$= \frac{1}{27} \begin{bmatrix} 2 & -1 \\ -1 & 2 \end{bmatrix}$$

$$\therefore \left\{ \begin{matrix} R_2 \\ R_3 \end{matrix} \right\} = - \frac{1}{27} \begin{bmatrix} 2 & -1 \\ -1 & 2 \end{bmatrix} \left\{ \begin{matrix} -10L \\ -8L \end{matrix} \right\} [w_1]$$

$$= -\frac{1}{27} \left\{ \begin{matrix} -12L \\ -6L \end{matrix} \right\} [w_1]$$

$$= \frac{1}{9} \left\{ \begin{matrix} 4L \\ 2L \end{matrix} \right\} [w_1] = \left\{ \begin{matrix} \frac{4}{9}Lw_1 \\ \frac{2}{9}Lw_1 \end{matrix} \right\}$$

$$\therefore \Delta_1 = \frac{L}{18EI} \begin{bmatrix} 8L^2 & -10L & -8L \end{bmatrix} \left\{ \begin{matrix} w_1 \\ \frac{4}{9}Lw_1 \\ \frac{2}{9}Lw_1 \end{matrix} \right\}$$

$$\quad\quad\quad\quad\quad 1\times 3 \quad\quad 3\times 1$$

$$\therefore \quad \Delta_1 = \frac{L}{18EI}\left[8L^2 w_1 - 10L\left(\frac{4}{9}Lw_1\right) - 8L\left(\frac{2}{9}Lw_1\right)\right]$$

$$= \frac{L}{18EI}\left(8L^2 w_1 - \frac{40}{9}L^2 w_1 - \frac{16 L^2 w_1}{9}\right)$$

$$= \frac{L}{18EI} \cdot \frac{(72 - 40 - 16)\, L^2 w_1}{9}$$

$$= \frac{L}{18EI} \cdot \frac{\overset{8}{\cancel{16}}\, L^2 w_1}{\underset{9}{9}}$$

$$\therefore \boxed{\Delta_1 = \frac{8\, w_1 L^3}{81 EI}}$$

the internal moments are given by:

$$\begin{Bmatrix} q_1 \\ q_2 \\ q_3 \\ q_4 \end{Bmatrix} = \begin{bmatrix} 0 & -1 & 0 \\ \dfrac{2L}{3} & -2/3 & -1/3 \\ \dfrac{2L}{3} & -2/3 & -1/3 \\ 0 & 0 & -1 \end{bmatrix} \begin{Bmatrix} w_1 \\ \dfrac{4Lw_1}{9} \\ \dfrac{2Lw_1}{9} \end{Bmatrix}$$

$$\qquad\qquad\qquad 4\times 3 \qquad 3\times 1$$

$$= \begin{Bmatrix} -\dfrac{4Lw_1}{9} \\[4pt] \dfrac{8}{27}Lw_1 \\[4pt] \dfrac{8}{27}Lw_1 \\[4pt] -\dfrac{2Lw}{9} \end{Bmatrix} = \frac{1}{27}\begin{Bmatrix} -12L \\ 8L \\ 8L \\ -6L \end{Bmatrix}[w_1]$$

* Steps to Analyze Indeterminate Structures by the
Flexibility-Matrix Method:

(1) Decide on the degree of indeterminacy and remove a sufficient number of redundants to form a determinate structure.

(2) To the determinate structure apply the known loads $\{W\}$ and the unknown redundants $\{R\}$. These forces together with their corresponding displacements form the structure loads and displacements.

(3) Subdivide the structure into elements and define the element forces $\{q\}$ and the corresponding displacements $\{\delta\}$.

(4) Form the composite element-flexibility matrix $[f_c]$ defined by the relation:

$$\{\delta\} = [f_c]\{q\}$$

(5) Form the force-transformation matrix $[B]$ for the determinate structure, and partition it into two submatrices corresponding to the known loads $\{W\}$ and the unknowns $\{R\}$. Thus:

$$\{q\} = [B_1 \mid B_2]\left\{\frac{W}{R}\right\}$$

(6) Form the structure-flexibility matrix $[F]$ for the determinate structure using the relation:

$$[F] = [B]^T [f_c][B] \quad , \quad \text{and partition } [F]$$

into four submatrices. Thus:

$$\left\{\frac{\Delta_W}{\Delta_R}\right\} = \left[\begin{array}{c|c} F_{11} & F_{12} \\ \hline F_{21} & F_{22} \end{array}\right]\left\{\frac{W}{R}\right\}$$

<u>Note</u>: The horizontal partitioning of $[F]$ corresponds to the separation of $\{\Delta\}$ into the unknown deflections $\{\Delta_W\}$ and the known deflections $\{\Delta_R\}$. Similarly, the vertical partitioning of $[F]$ corresponds to the separation of the loads into $\{W\}$ and $\{R\}$.

Note: It is easier for computer programming to obtain the four submatrices of $[F]$ individually (no need to obtain $[F]$ as a single matrix):

$$[F] = [B]^T [f_c] [B] = \begin{bmatrix} B_1^T \\ \hline B_2^T \end{bmatrix} [f_c] [B_1 \mid B_2]$$

$$\begin{bmatrix} F_{11} & \vdots & F_{12} \\ \hline F_{21} & \vdots & F_{22} \end{bmatrix} = \begin{bmatrix} B_1^T \\ \hline B_2^T \end{bmatrix} [f_c] [B_1 \mid B_2]$$

$$\implies [F_{11}] = [B_1^T][f_c][B_1]$$

$$[F_{12}] = [B_1^T][f_c][B_2]$$

$$[F_{21}] = [B_2^T][f_c][B_1]$$

$$[F_{22}] = [B_2^T][f_c][B_2]$$

$$\implies \left\{ \frac{\Delta_w}{\Delta_R} \right\} = \begin{bmatrix} B_1^T f_c B_1 & \vdots & B_1^T f_c B_2 \\ \hline B_2^T f_c B_1 & \vdots & B_2^T f_c B_2 \end{bmatrix} \left\{ \frac{w}{R} \right\}$$

(7) Solve the lower part of the above equation for $\{R\}$:

set $\{\Delta_R\} = \{0\}$ (zero vector).

$$\{\Delta_w\} = \begin{bmatrix} B_1^T f_c B_1 & \vdots & B_1^T f_c B_2 \end{bmatrix} \left\{ \frac{w}{R} \right\}$$

$$\{\Delta_R\} = \begin{bmatrix} B_2^T f_c B_1 & \vdots & B_2^T f_c B_2 \end{bmatrix} \left\{ \frac{w}{R} \right\}$$

$$\{0\} = \begin{bmatrix} F_{21} & \vdots & F_{22} \end{bmatrix} \left\{ \frac{w}{R} \right\}$$

$$\{0\} = [F_{21}]\{w\} + [F_{22}]\{R\}$$

$$[F_{22}]\{R\} = -[F_{2,}]\{w\}$$

$$\boxed{\{R\} = -[F_{22}]^{-1}[F_{2,}]\{w\}} \qquad \text{(solve simultaneous equations).}$$

(8) Substitute $\{R\}$ in $\{\Delta w\}$:

$$\{\Delta w\} = [F_{,,} \mid F_{12}]\left\{\frac{w}{R}\right\} = [F_{,,} \mid F_{12}]\left\{\frac{w}{-[F_{22}]^{-1}[F_{2,}]\{w\}}\right\}$$

$$\therefore \{\Delta w\} = [F_{,,}]\{w\} - [F_{12}][F_{22}]^{-1}[F_{2,}]\{w\}$$

$$\boxed{\{\Delta w\} = \left([F_{,,}] - [F_{12}][F_{22}]^{-1}[F_{2,}]\right)\{w\}}$$

(9) Obtain the internal forces $\{q\}$.

$$\{q\} = [B_, \mid B_2]\left\{\frac{w}{R}\right\} = [B_, \mid B_2]\left\{\frac{w}{-[F_{22}]^{-1}[F_{2,}]\{w\}}\right\}$$

$$= [B_,]\{w\} - [B_2][F_{22}]^{-1}[F_{2,}]\{w\}$$

$$\boxed{\{q\} = \left([B_,] - [B_2][F_{22}]^{-1}[F_{2,}]\right)\{w\}}$$

Note:
$$\{q\} = [\bar{B}]\{w\}$$

$$\text{where} \quad [\bar{B}] = [B_,] - [B_2][F_{22}]^{-1}[F_{2,}]$$

modified force-transformation for indeterminate structures

$$\{\Delta w\} = [\bar{F}]\{w\}$$

$$\text{where} \quad [\bar{F}] = [F_{12}][F_{22}]^{-1}[F_{2,}]$$

modified structure-flexibility matrix for indeterminate structures.

24

Homework 8

15.4 , 15.10 , 15.12 .

Influence Lines for Indeterminate Structures:

* The main purpose of influence lines for both statically determinate and indeterminate structures is to determine where to position moving loads to cause the maximum effect of a design function.

* Two approaches are presented for influence line construction:

① Quantitative Influence Lines : using the prior methods for structural analysis

② Qualitative Influence Lines : using the Müller – Breslau principle.

---

① Quantitative Influence Lines :

Example :

By the method of consistent deformations :

$$\delta_{Bo} + R_B f_{BB} = 0$$

$$\Rightarrow \boxed{R_B = \frac{-\delta_{Bo}}{f_{BB}}}$$

If the influence line for R_B is required, the real load on the beam is a moving unit concentrated load .

Thus $\delta_{Bo} = \delta_{Box}$

δ_{Bo} : deflection at B due to a unit load at x .

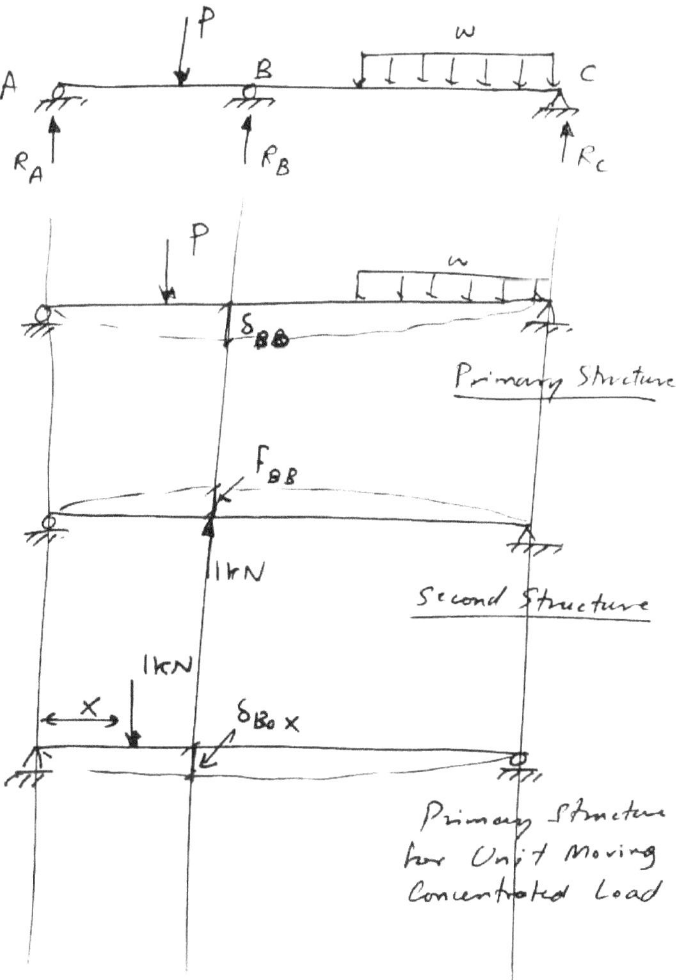

But $\quad S_{Box} = S_{XBo}$ (Maxwell's Law of Reciprocal $\quad$ [A2

Deflections).

where S_{XBo}: deflection at point x due to a unit load at B

$$S_{Box} + R_B f_{BB} = 0$$

$$S_{XBo} + R_B f_{BB} = 0$$

$$\Rightarrow \quad \boxed{R_B = -\frac{S_{XBo}}{f_{BB}}}$$

* The influence line for R_B is found by placing the unit load on the determinate (primary) beam at point B, followed by computation of the deflection at several x-beam locations including point B.

<u>Note</u>: It is observed that influence lines for indeterminate beams are <u>nonlinear</u> functions since they depend on deflection methods for their solution

---

<u>Example 1</u>:

Determine the <u>ordinate</u> values at 5-ft intervals and draw the influence lines for all support reactions and for shear and bending moment at point D on the continuous beam shown in the figure. $EI = $ constant.

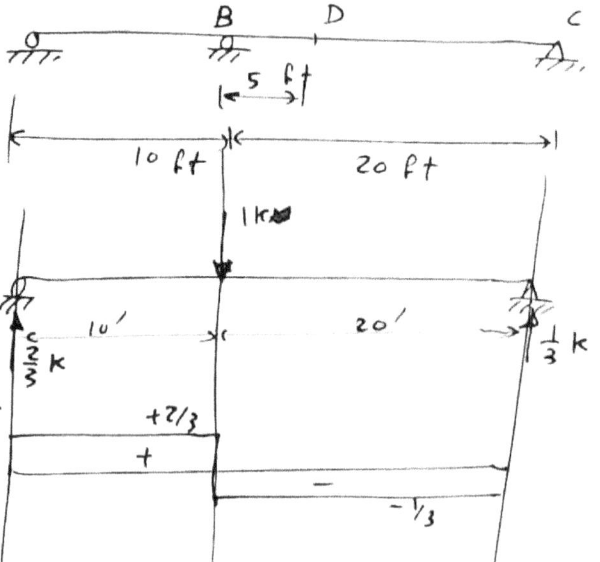

<u>Solution</u>:

Remove the redundant, R_B.

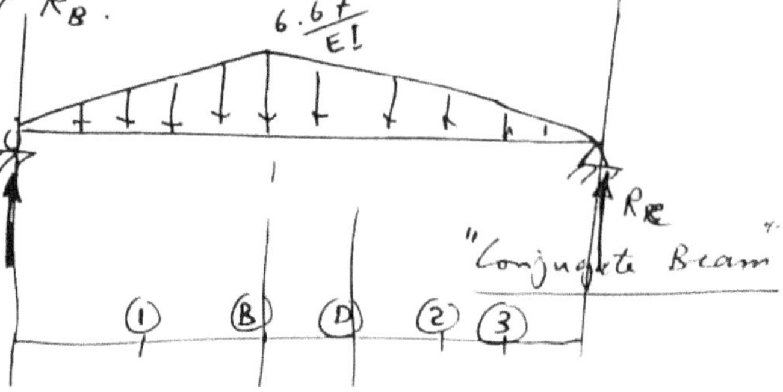

Divide the beam into 5-ft intervals as shown in the figure.

For the conjugate beam:

$+\circlearrowleft \; \Sigma m_{C} = 0:$

$$-R_{A}(30) + \frac{1}{2}\left(\frac{6.67}{EI}\right)(10)\left(20 + \frac{10}{3}\right) + \frac{1}{2}\left(\frac{6.67}{EI}\right)(20)\left(\frac{2}{3}\right)(20) = 0$$

$$\Rightarrow \; R_{A} = \frac{55.58}{EI} \uparrow$$

$+\uparrow \; \Sigma F_{y} = 0. \quad \frac{55.58}{EI} - \frac{1}{2}\left(\frac{6.67}{EI}\right)(30) \qquad + R_{C} = 0$

$$\Rightarrow \; R_{C} = \frac{44.47}{EI} \uparrow$$

$$\delta_{x_{1}} = \frac{55.58}{EI}(5) - \frac{1}{2}\left(\frac{3.33}{EI}\right)(5)\left(\frac{5}{3}\right)$$

$$= \frac{264}{EI}$$

$$\delta_{x_{B}} = \frac{55.58}{EI}(10) - \frac{1}{2}\left(\frac{6.67}{EI}\right)(10)\left(\frac{10}{3}\right)$$

$$= \frac{445}{EI}$$

$$\delta_{x_{D}} = \frac{44.47}{EI}(15) - \frac{1}{2}\left(\frac{5}{EI}\right)(15)\left(\frac{15}{3}\right) = \frac{480}{EI}$$

$$\delta_{x_{2}} = \frac{44.47}{EI}(10) - \frac{1}{2}\left(\frac{3.33}{EI}\right)(10)\left(\frac{10}{3}\right) = \frac{389}{EI}$$

$$\delta_{x_{3}} = \frac{44.47}{EI}(5) - \frac{1}{2}\left(\frac{1.67}{EI}\right)(5)\left(\frac{5}{3}\right) = \frac{215}{EI}$$

Note that $\quad f_{BB} = -\delta_{x_{B}} = -\frac{445}{EI}$

$$\Rightarrow \quad R_{B} = -\frac{\delta_{x_{B0}}}{f_{BB}} = \;\cdots (?) \quad \text{for each value } \uparrow \text{ the above.}$$

At point 1: $\quad R_{B} = -\frac{264}{-445} = 0.593$

At point B: $\quad R_{B} = -\frac{445}{-445} = 1.0$

At point D: $\quad R_{B} = -\frac{480}{-445} = 1.079$

At point 2 : $R_B = -\dfrac{389}{-445} = 0.874$

At point 3. $R_B = -\dfrac{215}{-445} = 0.483$

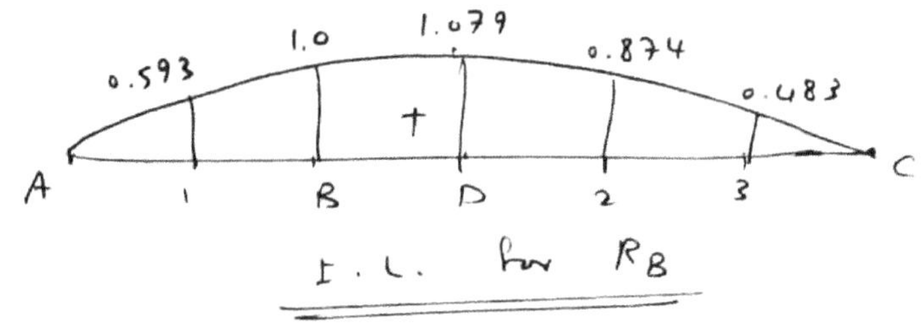

To find I.L. for R_A.

$+\circlearrowleft \Sigma M_c = 0:$

$1(30-x) - R_B(20) - R_A(30) = 0$

$30 R_A + 20 R_B = 30 - x$

$$R_A = \frac{30 - x - 20 R_B}{30}$$

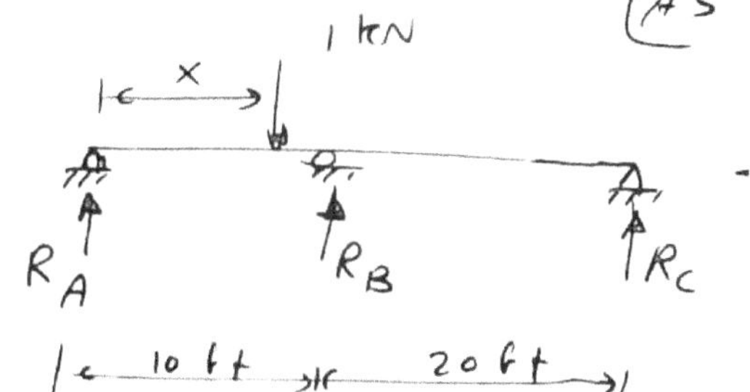

To find I.L. for R_c.

$\uparrow \Sigma F_y = 0:$ $R_A + R_B + R_c - 1 = 0$

$$R_c = 1 - R_A - R_B$$

x	R_B	R_A	R_c
0	0.0	1.0	0.0
5	0.593	0.438	-0.031
10	1.0	0.0	0.0
15	1.08	-0.22	0.14
20	0.874	-0.249	0.375
25	0.483	-0.155	0.672
30	0.0	0.0	1.0

* Determine the influence lines for the <u>shear</u> and <u>moment</u> at point D.

<u>Case (1)</u> : Unit load to the left of D:

$+\uparrow \Sigma F_y = 0:$

$$V_D + R_C = 0$$

$$\Rightarrow \boxed{V_D = -R_C}$$

$+\circlearrowleft \Sigma M_D = 0.$

$$-M_D + R_C(15) = 0$$

$$\boxed{M_D = 15 R_C}$$

<u>Case (2)</u> : Unit load to the right of D:

$+\uparrow \Sigma F_y = 0:$

$$R_A + R_B - V_D = 0$$

$$\Rightarrow \boxed{V_D = R_A + R_B}$$

$+\circlearrowleft \Sigma M_D = 0:$

$$M_D - R_B(5) - R_A(15) = 0$$

$$\Rightarrow \boxed{M_D = 15 R_A + 5 R_B}$$

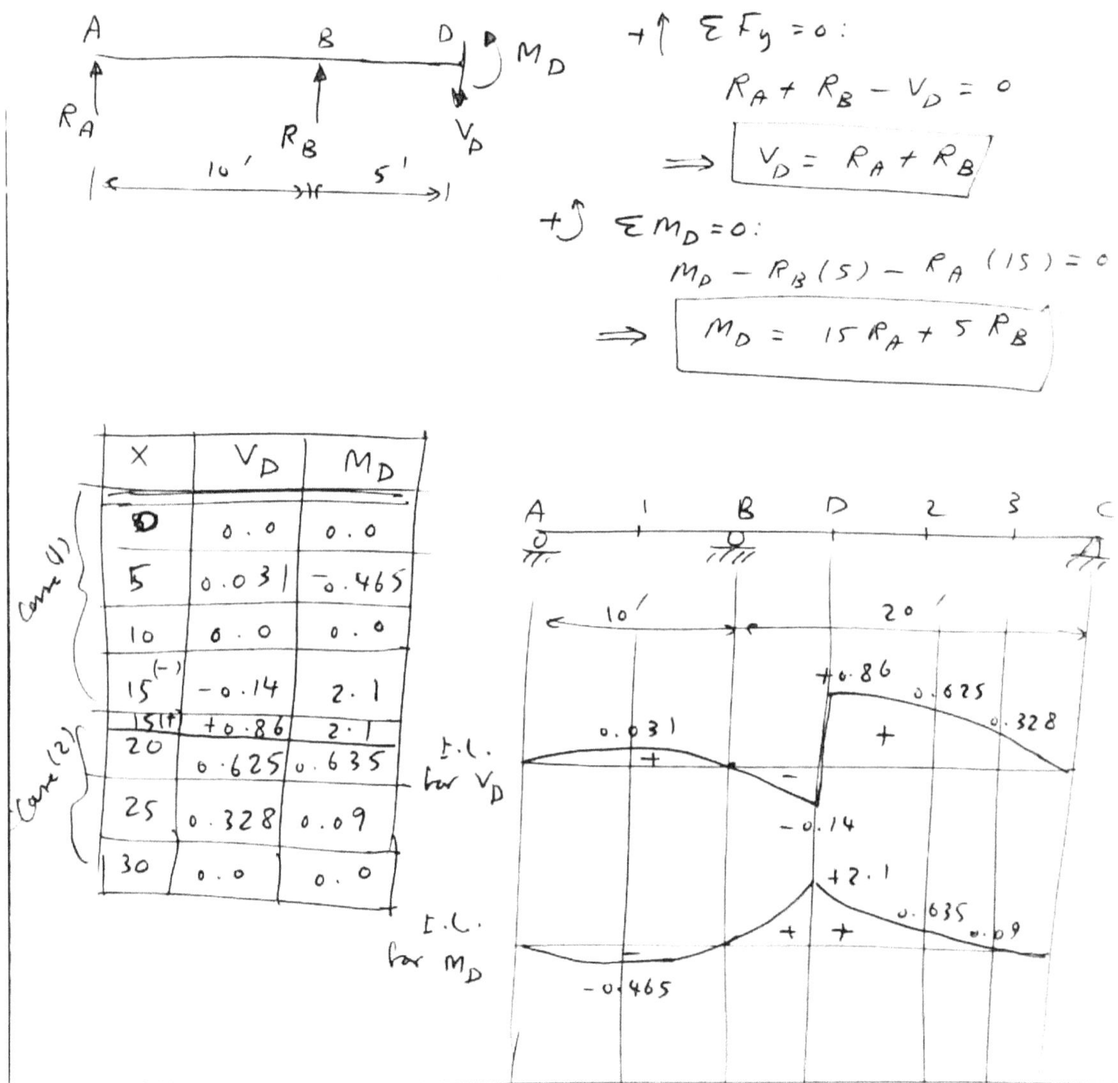

X	V_D	M_D
D	0.0	0.0
5	0.031	−0.465
10	0.0	0.0
$15^{(-)}$	−0.14	2.1
$15^{(+)}$	+0.86	2.1
20	0.625	0.635
25	0.328	0.09
30	0.0	0.0

Case (1) — rows D, 5, 10, 15
Case (2) — rows 15, 20, 25, 30

I.L. for V_D

I.L. for M_D

Qualitative Influence Lines — The Müeller–Breslau Principle:

* The importance of the influence line is fully realized when an analyst must determine where to position design live loads on a highly indeterminate structure that will maximize a design function.

* The design function may be shear, bending moment, axial force, slope change, deflection, etc.

* The structure in question may be a multi-story building, a bridge girder resting on many supports, or a highly redundant truss.

* In any case, the analyst must consider where the live loads are to be placed, regardless of whether the analysis is performed by classical hand-calculation methods or by use of the computer.

* the Müeller–Breslau Principle:

> If a function at a point on a structure, such as reaction, or shear, or moment, is allowed to act without restraint, the deflected shape of the structure, to some scale, represents the influence line of the function.

* This is significant since the major reason for drawing influence lines is to locate critical live-load positions for a particular function.

Example:

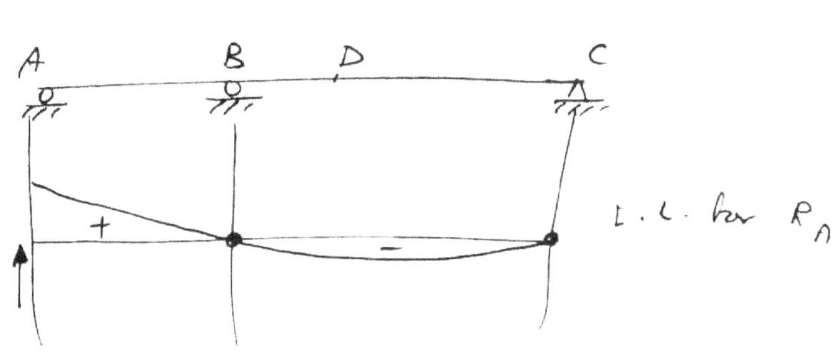

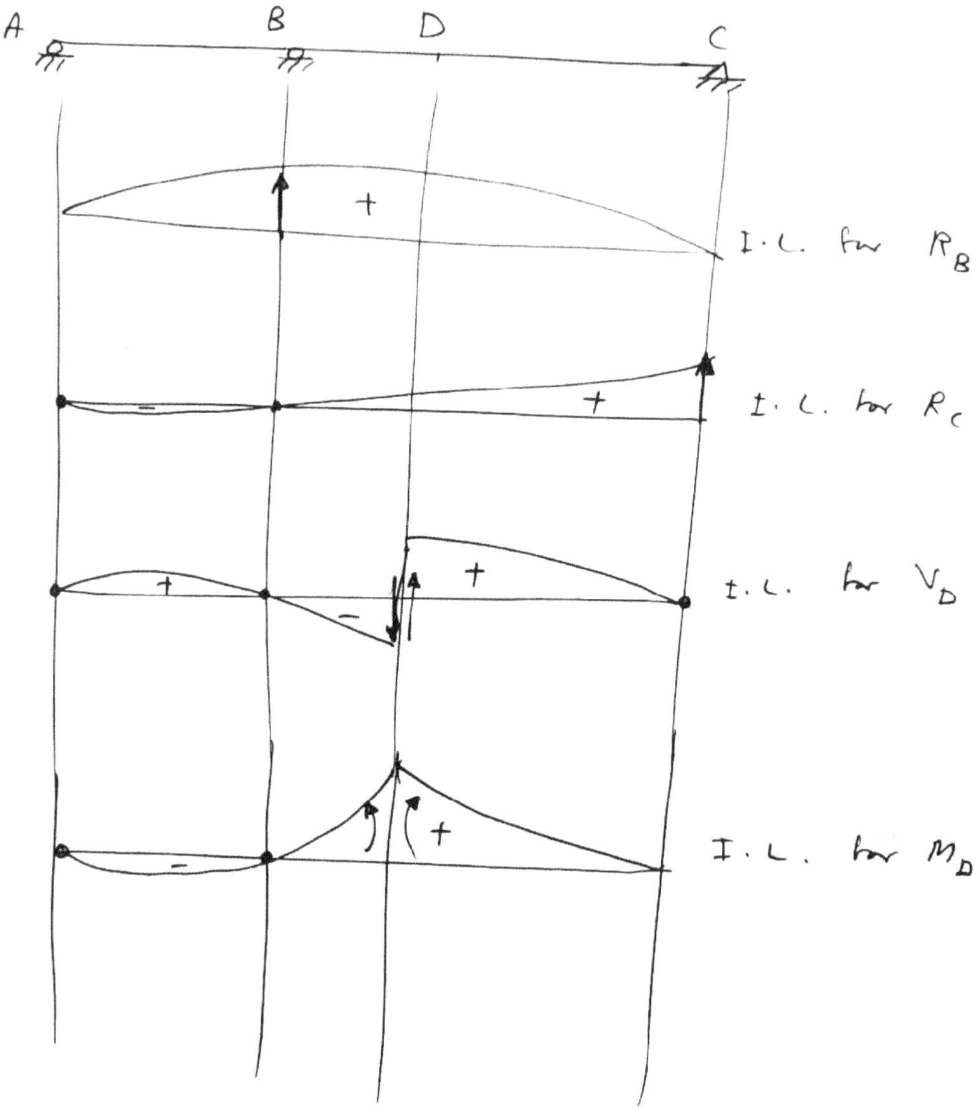

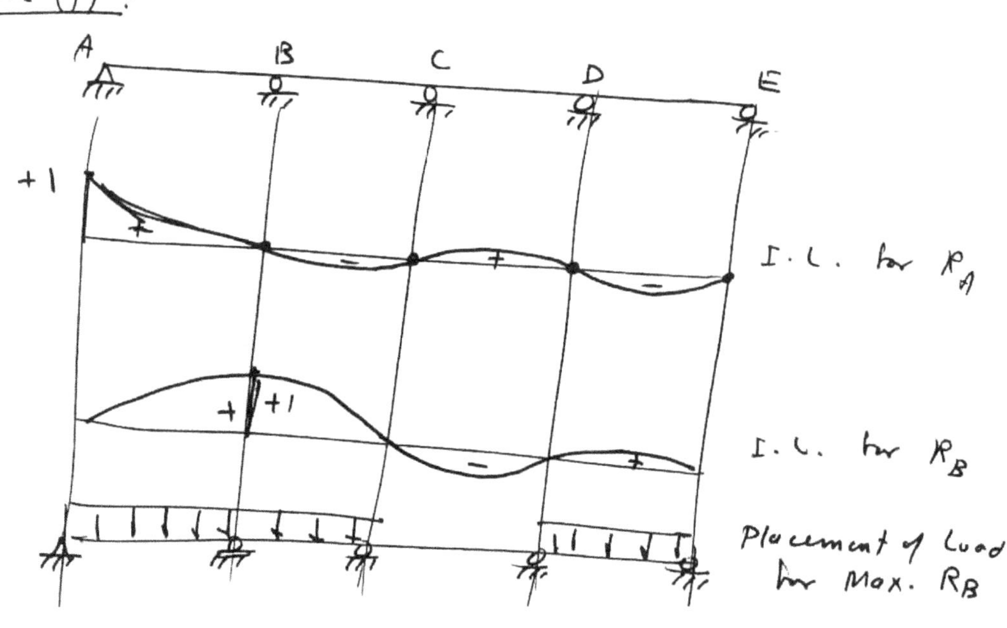

Critical Live-Load Placement:

Example (1):

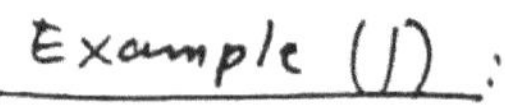

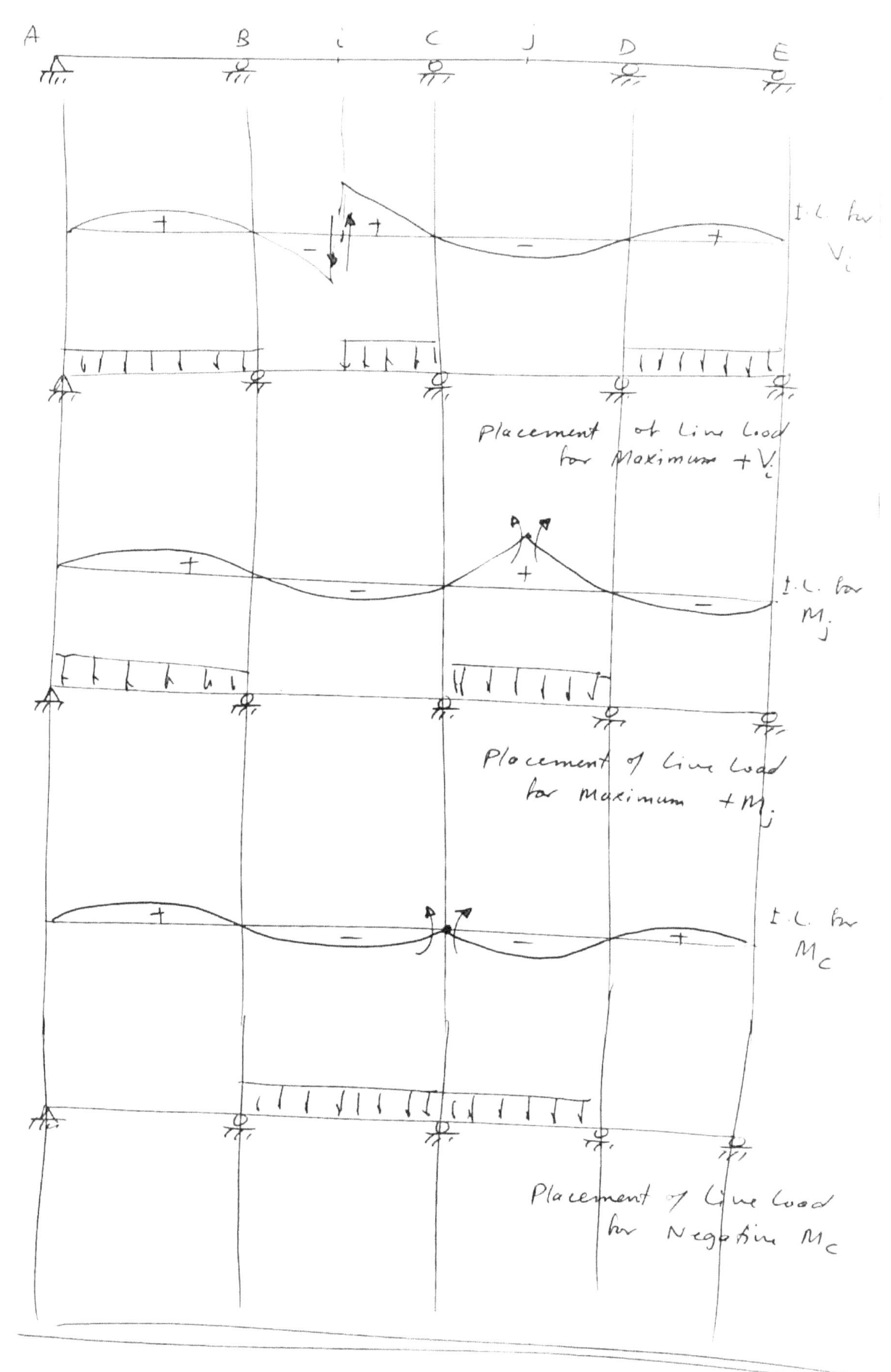

A   B   i   C   J   D   E
I.L. for V_i
Placement of Live Load for Maximum +V_i
I.L. for M_j
Placement of Live Load for Maximum +M_j
I.L. for M_c
Placement of Live Load for Negative M_c

$$\underline{Appendix}$$
$$\underline{Matrix\ Algebra}$$

$\underline{Definitions}$:

* A $\underline{matrix}$ is a rectangular array of numbers, with m rows and n columns.

$$[A] = \begin{bmatrix} a_{11} & a_{12} & a_{13} \\ a_{21} & a_{22} & a_{23} \\ a_{31} & a_{32} & a_{33} \end{bmatrix}$$

$a_{11}, a_{12}, \dots , a_{33}$ are called the $\underline{elements}$ of the matrix.

$\Rightarrow$ a_{ij} : element at row i and column j of the matrix.

$\qquad$ (i, j are called $\underline{subscripts}$)

* The $\underline{order}$ or $\underline{size}$ of a matrix refers to the number of rows and columns that the matrix possesses. A matrix with m rows and n columns is said to be of order $m \times n$.

$\underline{Types\ of\ Matrices}$:

(1) $\underline{Row\ Matrix}$: $\qquad [B] = \begin{bmatrix} 1 & 2 & 7 & 4 \end{bmatrix}$

$\qquad$ ($\underline{a\ vector}$).

(2) $\underline{Column\ Matrix}$: $\qquad [C] = \begin{bmatrix} 5 \\ 2 \\ -4 \end{bmatrix} \equiv \begin{Bmatrix} 5 \\ 2 \\ -4 \end{Bmatrix}$

$\qquad$ ($\underline{a\ vector}$)

(3) $\underline{Square\ Matrix}$: $\qquad [A] = \begin{bmatrix} 4 & 2 & 1 \\ 6 & 0 & 2 \\ 2 & -1 & 12 \end{bmatrix}$

(4) $\underline{Diagonal\ Matrix}$:

$\underline{\text{Note}}$: The diagonal matrix is a square matrix with zeroes off the diagonal.

$$[E] = \begin{bmatrix} 2 & 0 & 0 \\ 0 & 1 & 0 \\ 0 & 0 & 7 \end{bmatrix}$$

main diagonal

(5) <u>Symmetric Matrix</u> :

A symmetric matrix is a square matrix whose elements are symmetric about the main diagonal.

$\Rightarrow$ $a_{ij} = a_{ji}$ for a symmetric matrix.

$$[F] = \begin{bmatrix} 3 & 7 & -2 \\ 7 & 15 & 4 \\ -2 & 4 & 10 \end{bmatrix}$$

(6) <u>Null Matrix</u> :

the null matrix is a matrix all of whose elements are zero (<u>zero matrix</u>)

$$[G] = \begin{bmatrix} 0 & 0 & 0 \\ 0 & 0 & 0 \\ 0 & 0 & 0 \end{bmatrix}$$

(7) <u>Identity Matrix</u> :

the identity matrix is a diagonal matrix all of whose elements are ones.

$\Rightarrow$ It is usually denoted by $[I]$.

$$[I] = \begin{bmatrix} 1 & 0 & 0 \\ 0 & 1 & 0 \\ 0 & 0 & 1 \end{bmatrix}$$

<u>Equality, Addition and Subtraction</u> :

<u>Equality</u> : For two matrices to be equal they must be identical term by term.

$\Rightarrow$ Two equal matrices will necessarily be of the same order.

$$[A] = \begin{bmatrix} 3 & 6 \\ 4 & -2 \\ 1 & 7 \end{bmatrix} \quad , \quad [B] = \begin{bmatrix} 3 & 6 \\ 4 & -2 \\ 1 & 7 \end{bmatrix}$$

$\Rightarrow$ $[A] = [B]$

Addition and Subtraction:

Addition or subtraction of matrices is possible only if the matrices are of the same order.

$$\begin{bmatrix} 3 & 1 & 4 \\ 6 & -7 & 2 \end{bmatrix} + \begin{bmatrix} 2 & 0 & -1 \\ 6 & 2 & 1 \end{bmatrix} = \begin{bmatrix} 5 & 1 & 3 \\ 12 & -5 & 3 \end{bmatrix}$$

Multiplication:

(1) Multiplication of a **scalar** by a matrix.
   (**Scalar Multiplication**)

$$3 \begin{bmatrix} 2 & 4 \\ 6 & 2 \end{bmatrix} = \begin{bmatrix} 6 & 12 \\ 18 & 6 \end{bmatrix}$$

scalar ↗   matrix ↗

(2) Multiplication of Two Matrices:
   (**Matrix Multiplication**)

$$a_{11} x_1 + a_{12} x_2 = c_1$$

$$a_{21} x_1 + a_{22} x_2 = c_2$$

let $[A] = \begin{bmatrix} a_{11} & a_{12} \\ a_{21} & a_{22} \end{bmatrix}$, $\{X\} = \begin{Bmatrix} x_1 \\ x_2 \end{Bmatrix}$, $\{C\} = \begin{Bmatrix} c_1 \\ c_2 \end{Bmatrix}$

$$\Rightarrow \begin{bmatrix} a_{11} & a_{12} \\ a_{21} & a_{22} \end{bmatrix} \begin{Bmatrix} x_1 \\ x_2 \end{Bmatrix} = \begin{Bmatrix} c_1 \\ c_2 \end{Bmatrix}$$

$$\Rightarrow [A]\{X\} = \{C\} \Rightarrow \text{"matrix equation"}$$

c_{ij} is obtained by multiplying row i of $[A]$ by column j of $\{X\}$ term by term and adding the resulting products.

$$c_1 = a_{11} x_1 + a_{12} x_2$$

$$c_2 = a_{21} x_1 + a_{22} x_2$$

In general, $\quad [A][B] = [C]$

$$c_{ij} = \sum_{k=1}^{n} a_{ik} b_{kj}$$

where $\quad n =$ number of columns in A
$\qquad\quad = $ number of rows in B.

Example:
$$\begin{bmatrix} a_{11} & a_{12} & a_{13} \\ a_{21} & a_{22} & a_{23} \\ a_{31} & a_{32} & a_{33} \end{bmatrix} \begin{bmatrix} b_{11} & b_{12} & b_{13} \\ b_{21} & b_{22} & b_{23} \\ b_{31} & b_{32} & b_{33} \end{bmatrix} = \begin{bmatrix} c_{11} & c_{12} & c_{13} \\ c_{21} & c_{22} & c_{23} \\ c_{31} & c_{32} & c_{33} \end{bmatrix}$$

$$c_{11} = a_{11} b_{11} + a_{12} b_{21} + a_{13} b_{31} = \sum_{k=1}^{n} a_{1k} b_{k1}$$

$$c_{12} = a_{11} b_{12} + a_{12} b_{22} + a_{13} b_{32} = \sum_{k=1}^{n} a_{1k} b_{k2}$$

$$\vdots$$

$$c_{33} = a_{31} b_{13} + a_{32} b_{23} + a_{33} b_{33} = \sum_{k=1}^{n} a_{3k} b_{k3}$$

Example:

$$\begin{bmatrix} 1 & 3 & 2 \\ 0 & -2 & 5 \\ 4 & 1 & -3 \end{bmatrix} \begin{bmatrix} 3 & 1 \\ 0 & 4 \\ -1 & 2 \end{bmatrix} = \begin{bmatrix} 3+0-2 & 1+12+4 \\ 0+0-5 & 0-8+10 \\ 12+0+3 & 4+4-6 \end{bmatrix}$$

$\quad 3 \times \boxed{3} \qquad\qquad \boxed{3} \times 2$
$\qquad\qquad 3 \times 2$

$$= \begin{bmatrix} 1 & 17 \\ -5 & 2 \\ 15 & 2 \end{bmatrix}$$

---

* Matrix multiplication is associative and distributive.

<u>Associative</u>: $\quad [A][BC] = [AB][C]$

<u>Distributive</u>: $\quad [A][B+C] = [AB] + [AC]$

* Matrix multiplication is <u>not</u> <u>commutative</u>:

$$[A][B] \neq [B][A]$$

the only exception is when multiplying by the identity matrix.

$$\begin{pmatrix} 2 & 3 \\ 2 & 4 \end{pmatrix} \begin{pmatrix} 1 & 0 \\ 0 & 1 \end{pmatrix} = \begin{pmatrix} 2 & 3 \\ 2 & 4 \end{pmatrix}$$

$$\begin{pmatrix} 1 & 0 \\ 0 & 1 \end{pmatrix} \begin{bmatrix} 2 & 3 \\ 2 & 4 \end{bmatrix} = \begin{bmatrix} 2 & 3 \\ 2 & 4 \end{bmatrix}$$

$$\Rightarrow \quad [A][I] = [I][A] = [A].$$

Transpose of a Matrix :

* The __transpose__ of a matrix is obtained by interchanging the rows and columns of the matrix.

$$[A] = \begin{bmatrix} 3 & 2 & -1 \\ 6 & 2 & 4 \end{bmatrix} \qquad \underline{\underline{2 \times 3}}$$

the transpose of $[A]$ $\Rightarrow$ $[A]^T = \begin{bmatrix} 3 & 6 \\ 2 & 2 \\ -1 & 4 \end{bmatrix}$ $\qquad \underline{\underline{3 \times 2}}$

Example : Work = Force $\times$ Distance
(scalar)   (vector)   (vector)

$$\{F\} = \begin{Bmatrix} F_x \\ F_y \\ F_z \end{Bmatrix} \quad , \quad \{D\} = \begin{Bmatrix} D_x \\ D_y \\ D_z \end{Bmatrix}$$

$\qquad\qquad 3 \times 1 \qquad\qquad\qquad 3 \times 1$

Form the transpose of $\{F\}$:

$$W = \{F\}^T \{D\} = \begin{pmatrix} F_x & F_y & F_z \end{pmatrix} \begin{Bmatrix} D_x \\ D_y \\ D_z \end{Bmatrix}$$

$\qquad\qquad\qquad\qquad 1 \times 3 \qquad\quad 3 \times 1$

$$= \left(F_x D_x + F_y D_y + F_z D_z \right) \quad \underline{\underline{scalar}}$$

Determinants :

* the __determinant__ consists of a square array of numbers bounded by two vertical lines.

$$\begin{vmatrix} a & b \\ c & d \end{vmatrix} = ad - bc \qquad \text{For a } 2\times 2 \text{ matrix.}$$

* A determinant is a __scalar__ .

Laplace Expansion of Determinants .

__Minors__ :

$$|A| = \begin{vmatrix} a_{11} & a_{12} & a_{13} \\ a_{21} & a_{22} & a_{23} \\ a_{31} & a_{32} & a_{33} \end{vmatrix}$$

the minor of element a_{11} is $M_{11} = \begin{vmatrix} a_{22} & a_{23} \\ a_{32} & a_{33} \end{vmatrix}$

the minor of element a_{12} is $M_{12} = \begin{vmatrix} a_{21} & a_{23} \\ a_{31} & a_{33} \end{vmatrix}$

__Cofactors__ : the __cofactor__ of an element in a determinant is the mior of the element multiplied by $(-1)^{i+j}$, where i and j are the row and column designation of the element.

$$C_{11} = (-1)^{1+1} M_{11} = + M_{11} = + \begin{vmatrix} a_{22} & a_{23} \\ a_{32} & a_{33} \end{vmatrix}$$

$$C_{12} = (-1)^{1+2} M_{12} = - M_{12} = - \begin{vmatrix} a_{21} & a_{23} \\ a_{31} & a_{33} \end{vmatrix}$$

---

* A __singular matrix__ has a zero determinant.
* A __non-singular matrix__ has a nonzero determinant.

Laplace Expansion of Determinants:

* The value of the determinant is equal to the sum of the products of the elements and their cofactors in any row or column of the determinant.

$$|A| = \sum_{j=1}^{n} a_{ij} C_{ij} \qquad , \text{ for any } i.$$

Example:
$$|A| = \begin{vmatrix} 1 & 2 & -4 \\ 3 & 1 & -2 \\ 4 & 2 & 3 \end{vmatrix}$$

$$|A| = (1)\begin{vmatrix} 1 & -2 \\ 2 & 3 \end{vmatrix} - (2)\begin{vmatrix} 3 & -2 \\ 4 & 3 \end{vmatrix} + (-4)\begin{vmatrix} 3 & 1 \\ 4 & 2 \end{vmatrix}$$

$$= (1)(3+4) - 2(9+8) + (-4)(2)$$

$$= 7 - 34 - 8$$

$$= -35$$

Inverse of a Matrix: (equivalent to division):

* Inverse of $[B]$ is $[B]^{-1}$ defined by:

$$[B][B]^{-1} = [I]$$
$$[B]^{-1}[B] = [I]$$

* Only a non-singular matrix has an inverse.

$$
\left.
\begin{array}{l}
a_{11}x_1 + a_{12}x_2 + a_{13}x_3 = c_1 \\
a_{21}x_1 + a_{22}x_2 + a_{23}x_3 = c_2 \\
a_{31}x_1 + a_{32}x_2 + a_{33}x_3 = c_3
\end{array}
\right\}
\begin{array}{l}
\text{solve for} \\
x_1, x_2, x_3.
\end{array}
$$

$$\Rightarrow \begin{bmatrix} a_{11} & a_{12} & a_{13} \\ a_{21} & a_{22} & a_{23} \\ a_{31} & a_{32} & a_{33} \end{bmatrix} \begin{Bmatrix} x_1 \\ x_2 \\ x_3 \end{Bmatrix} = \begin{Bmatrix} c_1 \\ c_2 \\ c_3 \end{Bmatrix}$$

$$[A][X] = [C]$$

where $[A]$: coefficient matrix

$[X]$: unknown vector

$[C]$: constants vector.

Pre-multiply the equation by $[A]^{-1}$:

$$[A]^{-1}[A][X] = [A]^{-1}[C]$$

$$[I][X] = [A]^{-1}[C]$$

$$\boxed{\{X\} = [A]^{-1}\{C\}}$$

How can we determine $[A]^{-1}$:

Evaluate the <u>cofactor</u> matrix :

$$[Cof.\ A] = \begin{bmatrix} C_{11} & C_{12} & C_{13} \\ C_{21} & C_{22} & C_{23} \\ C_{31} & C_{32} & C_{33} \end{bmatrix}$$

Form the <u>adjoint</u> matrix : (transpose of $[Cof.\ A]$)

$$[Adj\ A] = [Cof.\ A]^{T} = \begin{bmatrix} C_{11} & C_{21} & C_{31} \\ C_{12} & C_{22} & C_{32} \\ C_{13} & C_{23} & C_{33} \end{bmatrix}$$

$$\Rightarrow \quad \boxed{[A]^{-1} = \frac{1}{|A|}[Adj.\ A]}$$

<u>Example</u> : Solve the equations :

$$x_1 + x_2 + x_3 = 2$$
$$2x_1 - x_2 + x_3 = 5$$
$$x_1 + 2x_2 + 2x_3 = 3$$

$$\begin{bmatrix} 1 & 1 & 1 \\ 2 & -1 & 1 \\ 1 & 2 & 2 \end{bmatrix} \begin{Bmatrix} x_1 \\ x_2 \\ x_3 \end{Bmatrix} = \begin{Bmatrix} 2 \\ 5 \\ 3 \end{Bmatrix}$$

$$\Rightarrow \quad [A] = \begin{bmatrix} 1 & +1 & 1 \\ 2 & -1 & 1 \\ 1 & 2 & 2 \end{bmatrix}$$

$$[Cof.\ A] = \begin{bmatrix} -4 & -3 & 5 \\ 0 & 1 & -1 \\ 2 & 1 & -3 \end{bmatrix}$$

$$[Adj.\ A] = \begin{bmatrix} -4 & 0 & 2 \\ -3 & 1 & 1 \\ 5 & -1 & -3 \end{bmatrix}$$

$$|A| = 1(-2-2) - 1(4-1) + 1(4+1) = -4-3+5$$
$$= -2.$$

$$\Rightarrow \quad [A]^{-1} = \frac{1}{|A|}[Adj.\ A] = \frac{1}{-2}\begin{bmatrix} -4 & 0 & 2 \\ -3 & 1 & 1 \\ 5 & -1 & -3 \end{bmatrix}$$

$$= \begin{bmatrix} 2 & 0 & -1 \\ 1.5 & -0.5 & -0.5 \\ -2.5 & 0.5 & 1.5 \end{bmatrix}$$

$$\Rightarrow \quad \{x\} = [A]^{-1}\{C\} = \begin{bmatrix} 2 & 0 & -1 \\ 1.5 & -0.5 & -0.5 \\ -2.5 & 0.5 & 1.5 \end{bmatrix} \begin{Bmatrix} 2 \\ 5 \\ 3 \end{Bmatrix}$$

$$\Rightarrow \quad \begin{Bmatrix} x_1 \\ x_2 \\ x_3 \end{Bmatrix} = \begin{Bmatrix} 1 \\ -1 \\ 2 \end{Bmatrix}$$

Partioning of Matrices:

* We can subdivide a matrix into several parts called __submatrices__.

* This process is known as __partitioning__, and is accomplished by using horizontal and vertical lines:

$$[A] = \left[\begin{array}{cc|cc} 3 & 2 & 6 & 2 \\ 1 & 4 & 1 & 1 \\ 2 & 7 & 4 & 1 \\ 3 & 2 & 1 & 7 \end{array}\right] = \begin{bmatrix} A_{11} & A_{12} \\ A_{21} & A_{22} \end{bmatrix}$$

Matrix Algebra on Partitioned Matrices:

$$[A] = \begin{bmatrix} 3 & 2 & | & 1 \\ 1 & 4 & | & 2 \\ \hline 4 & 2 & | & 1 \end{bmatrix} = \begin{bmatrix} A_{11} & A_{12} \\ A_{21} & A_{22} \end{bmatrix}$$

$$[B] = \begin{bmatrix} 4 & 2 \\ 3 & 6 \\ \hline 1 & 2 \end{bmatrix} = \begin{bmatrix} B_{11} \\ B_{21} \end{bmatrix}$$

$$[A][B] = \begin{bmatrix} A_{11} & A_{12} \\ A_{21} & A_{22} \end{bmatrix} \begin{bmatrix} B_{11} \\ B_{21} \end{bmatrix} = \begin{bmatrix} C_{11} \\ C_{21} \end{bmatrix}$$

where
$$C_{11} = A_{11} B_{11} + A_{12} B_{21}$$
$$C_{21} = A_{21} B_{11} + A_{22} B_{21}$$

$$\Rightarrow \quad A_{11} B_{11} = \begin{bmatrix} 3 & 2 \\ 1 & 4 \end{bmatrix} \begin{bmatrix} 4 & 2 \\ 3 & 6 \end{bmatrix} = \begin{bmatrix} 18 & 18 \\ 16 & 26 \end{bmatrix}$$

$$A_{12} B_{21} = \begin{bmatrix} 1 \\ 2 \end{bmatrix} \begin{bmatrix} 1 & 2 \end{bmatrix} = \begin{bmatrix} 1 & 2 \\ 2 & 4 \end{bmatrix}$$
$$\underset{2\times 1}{} \qquad \underset{1\times 2}{}$$

$$A_{21} B_{11} = \begin{bmatrix} 4 & 2 \end{bmatrix} \begin{bmatrix} 4 & 2 \\ 3 & 6 \end{bmatrix} = \begin{bmatrix} 22 & 20 \end{bmatrix}$$
$$\underset{1\times 2}{} \qquad \underset{2\times 2}{}$$

$$A_{22} B_{21} = \begin{bmatrix} 1 \end{bmatrix} \begin{bmatrix} 1 & 2 \end{bmatrix} = \begin{bmatrix} 1 & 2 \end{bmatrix}$$
$$\underset{1\times 1}{} \qquad \underset{1\times 2}{}$$

$$\Rightarrow \quad C_{11} = \begin{bmatrix} 18 & 18 \\ 16 & 26 \end{bmatrix} + \begin{bmatrix} 1 & 2 \\ 2 & 4 \end{bmatrix} = \begin{bmatrix} 19 & 20 \\ 18 & 30 \end{bmatrix}$$

$$C_{21} = \begin{bmatrix} 22 & 20 \end{bmatrix} + \begin{bmatrix} 1 & 2 \end{bmatrix} = \begin{bmatrix} 23 & 22 \end{bmatrix}$$

$$[C] = \begin{bmatrix} C_{11} \\ C_{21} \end{bmatrix} = \begin{bmatrix} 19 & 20 \\ 18 & 30 \\ \hline 23 & 22 \end{bmatrix}$$

---

* For multiplication of partitioned matrices, the vertical partitioning of the first must be identical to the horizontal partitioning of the second.

www.ingramcontent.com/pod-product-compliance
Lightning Source LLC
Chambersburg PA
CBHW060559120726
48002CB00010B/2739